Signals and Information

C. C. Goodyear,

Colin Crosland

B.Sc., D.Phil., A.Inst.P.

WILEY—INTERSCIENCE

A Division of John Wiley & Sons, Inc.,

NEW YORK — LONDON — SYDNEY — TORONTO

First Published in 1971 by Butterworth & Co. (Publishers) (London).

Published in the U.S.A. by Wiley-Interscience Division,
John Wiley & Sons, Inc., 605 Third Avenue, New York, N.Y.10016
Library of Congress Catalog Card Number: 74-38234
Wiley ISBN: 0-471-31520-6

Printed in Hungary

Contents

Contents

3 SIGNAL PROCESSING

4 MODULATION THEORY

5 PROBABILITY AND STATISTICS

6 NOISE

7 SIGNALS WITH NOISE

Contents

Preface

This book is offered as an introduction, at final-year undergraduate level, to the theory of signals, signal processing, noise and information. Its main aim is to provide a background of communication theory for students who are specialising in telecommunications. I hope, however, that the treatment is sufficiently general for the material to be of help to students in any discipline where a knowledge of signal processing techniques is required. The emphasis is on fundamental theory and details of circuits and practical applications have been touched on only lightly. Here and there I have given references to places in the literature where descriptions of practical systems are to be found. These are not intended as any sort of guide to the literature, but as extensions and illustrations of some of the topics covered in the text.

Most students who are encountering this material for the first time will find that a few of the more mathematical sections take a little time to digest. I have therefore tried to group the material in such a way that the reader who prefers to skip over the more difficult sections at first reading, may do so without losing many of the major topics. For example, the reader may leave Chapter 2 at the point where the Fourier transform is introduced and still cope with the impulse and step response of idealised low-pass systems in Chapter 3, since these are first treated by a Fourier series method. He will have omitted the sampling theorem from Chapter 2, but the basic idea is repeated in Section 3.16. A suggested 'first reading' sequence would be the following: Chapter 1, Chapter 2 omitting Sections 2.11–2.18, Chapter 3 omitting

Preface

Sections 3.6, 3.7, 3.8, 3.11 and 3.12, Chapter 4 omitting Section 4.5, Chapters 6, 8 and 9.

There are one or two novelties but by and large the material is fairly standard, differing from other treatments in the order of presentation and in the relative proportions of the mixture of signal theory, optimum processing and information theory. I believe that the result is a well balanced and coherent course which may appeal to others who are involved with the subject as teachers.

Three of the figures are taken or adapted from other publications and the kindness of the authors concerned is acknowledged in the text. Much is owed to the clear accounts of several topics covered in existing texts and these are listed in the 'suggestions for further reading'. I have also derived considerable benefit from discussions with my colleagues Mr. J. Durnford and Dr. P. A. W. Walker on a wide range of topics over the years in which I have been teaching this course. My gratitude is extended to them and also to Professor J. M. Meek for his interest and encouragement.

C. C. GOODYEAR

Department of Electrical Engineering and Electronics University of Liverpool

Chapter 1

Introduction

1.1 SIGNALS

The term 'signal' is usually understood to mean a pattern of some kind which is used to convey a message or part of a message. We may think of the letters on a printed page in this way, or of the flags used for signalling at sea, or of the patterns of dots and dashes in the Morse code. The medium upon which the message pattern is impressed can, in almost all cases, support many more patterns than are required for the purposes of communication. Fig. 1.1, for example, shows a few of the patterns which may be made by blocking out any number of the squares of a 20×20 grid. The fraction of possible patterns which correspond exactly, as at (c), or even approximately, as at (d), to letters of the alphabet is clearly very small. Similarly the voltage difference between a pair of terminals may be made to fluctuate in a limitless variety of ways. The curves of Fig. 1.2 are plots of voltage against time which, apart from certain constraints (namely that there should be no fluctuations faster that a certain rate and that the mean square voltage should be about the same for each), have been drawn quite randomly with the exception of the last, which was made to resemble the 'dot dot dot dash' pattern for the letter V in Morse code. For the purposes of our discussion it will be helpful to use the term *signal* quite generally for any of the possible patterns of Figs 1.1 and 1.2. Of the huge variety of possible signals, a few have—or may be given by

1

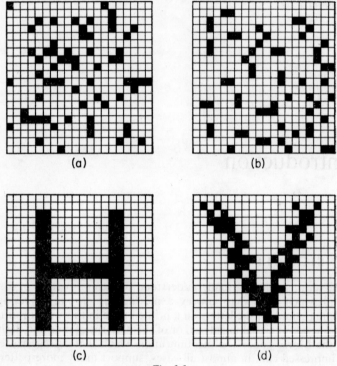

(a) (b)

(c) (d)

Fig. 1.1.

choice—a special significance as *message signals*. Thus Fig. 1.1 (c) would be classed as a message signal.

Signals in this general sense may be produced by quite random processes. Patterns like Fig. 1.1 can be constructed by tossing a penny to decide for each square in turn whether it is to be black or white. It can then be shown, from the theory of probability, that the chance of any of the patterns (a)–(d) turning up is exactly the same for each. From the statistical point of view there is nothing 'special' about message signals and when a message signal is received we cannot tell, by examining the signal alone, whether it originated at the transmitter as a 'true' message signal or whether it has resulted from random fluctuations somewhere in the system or whether it is some combination of the two.

We have all seen faces in the fire or heard whispers in the wind

2

and it is interesting to consider how this comes about. The glowing embers throw up a myriad of different images and the breeze calls out a whole company of sounds. Consciously or subconsciously our minds process these incoming signals and just a few evoke a special response. They are those that match a pattern which is already in the mind as a familiar form of 'message signal'; the remainder attract little attention. Notice that the match need not be exact, for the eye and ear are trained to accept not only a given message

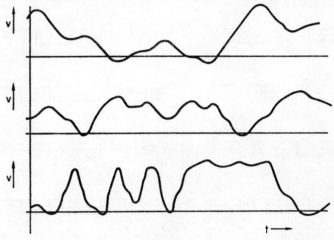

Fig. 1.2.

signal but also a range of signals lying close by, as the same thing. A mathematical account of what is meant by 'lying close by' must be delayed until much later in the book (mainly in Chapter 9), but for the present we may rely on the intuitive notion that Fig. 1.1 (d), for example, lies close enough to the letter v to be accepted as such.

The process of communication may be described in fairly general terms as follows. It is understood that some group of patterns which constitute the message signals has been agreed between the sender and the receiver. A message signal is transmitted and may become distorted and suffer the addition of random fluctuations to a greater or lesser degree before reaching the receiver. The fundamental problem at the receiver is to decide, on examining the incoming signal, which message signal was transmitted. The risk

3

of being deceived by a 'whisper in the wind' must be made as small as possible.

In this book we shall be largely concerned with electrical rather than with visual or other types of signal, though a television camera or facsimile machine could be used to convert from a signal like those of Fig. 1.1 to an electrical analogue signal. The signals which are carried by the cable from a microphone in a broadcasting studio are recognised as message signals by the listener, after they have been converted back into sound waves. On the other hand, the

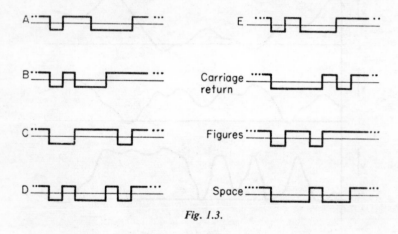

Fig. 1.3.

process of pattern recognition in some instances can be entirely electrical. Fig. 1.3 shows the signal patterns which are used to transmit some of the symbols on a teleprinter keyboard. The teleprinter code is basically a five-unit binary code, using current in one direction to represent a 1 and current in the reverse direction (or in some systems, zero current) to represent a 0. In transmission the five elements are preceded by a 'start' signal and followed by a 'stop' signal, as shown in Fig. 1.4. A teleprinter which is receiving a message remains in a quiescent state until the start signal arrives, after which it examines the incoming signal at five instants corresponding with the five code elements. The stop signal returns it to the quiescent state. The value of the signal voltage at the five observation times determines which letter is to be printed. Teleprinter code patterns are simple, voiced sound patterns are very complicated, but both are message signals which the receiver, human or machine, recognises. The process of recognition may be very complex but must essentially involve comparing the received signal with each expected possible message signal. The one selected is, ideally, the message signal which 'lies nearest', i.e. which shows the best correlation. As we shall see in a later chapter, sophisticated receivers can be designed which employ 'correlation detectors' and perform precisely this operation.

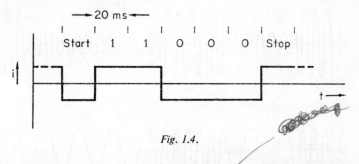

Fig. 1.4.

1.2 MODULATION AND SIGNAL PROCESSING

It is frequently inconvenient or impossible to transmit electrical message signals in their original 'raw' state. The pattern is then translated into a new form which is better suited to the communication channel available. Fig. 1.5(a) shows a signal which might be part of the output from a teleprinter. The curve (b) below it shows the same pattern now appearing in the envelope of a relatively high-frequency sinusoidal signal or 'carrier' wave. It clearly carries the same information as (a). Another method is shown at (c), where the amplitude of the carrier remains constant, but its frequency changes abruptly between two values, according to the state of signal (a). A third alternative is given at (d), in which the amplitude and frequency are fixed, but the phase changes by 180° at each jump in signal (a). Using appropriate circuits, signal (a) may be changed into any of these new forms, transmitted over the communication link, and then translated back again to the original pattern. The term *modulation* is used for a reversible process which changes the signal pattern into a new pattern carrying the same information.

A particularly interesting technique is shown in Fig. 1.6 in which the message waveform (a) modulates a pulse train carrier. At a succession of equally spaced moments in time, a pulse may or

5

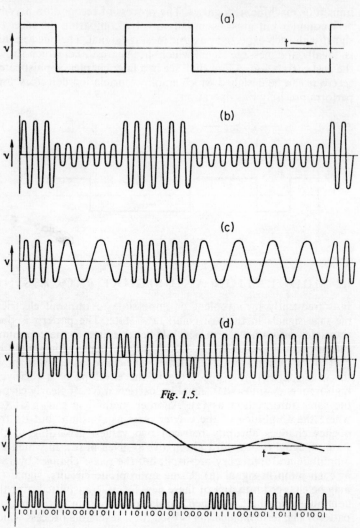

Fig. 1.5.

Fig. 1.6.

may not occur, the choice being determined in a way which will be explained later. The pulses which do occur are all of the same height and duration, and the pulse pattern can be represented as a se-

6

quence of 1's and 0's as shown at the bottom of the figure. This method of modulation (pulse code modulation) is ideally adapted to pulse circuit techniques and, with the advent of integrated circuits, will probably become more and more widely used.

It is at first astonishing to learn that a Beethoven symphony—or more strictly a particular performance of a Beethoven symphony—may be written as one long binary sequence. A fuller discussion of pulse code modulation and other modulation techniques is given below in Chapter 4, but it is first necessary to prepare the ground with an account of the mathematical analysis of signals, and this is the concern of Chapter 2. It is also important to study the extent to which signal patterns may become distorted in transmission. A certain amount of distortion is tolerable, but the signals may be distorted beyond recognition if, for example, the bandwith available for transmission is too narrow or too fast a signalling speed is attempted. Non-linear circuit elements also introduce distortion. Signal transmission in linear and non-linear systems and some related signal processing techniques are treated in Chapter 3.

1.3 NOISY SIGNALS

The background hiss which accompanies the speech or music from even the best-designed radio receiver is just one manifestation of the ubiquitous problem of noise. The effects of random electrical disturbances, picked up by the aerial or produced within active devices or even in a simple resistor, can be reduced by careful design but never totally eliminated. The message signal pattern is always received against a jumbled background. The noise background cannot be dismissed as an unpleasant nuisance which can be overcome simply by increasing the transmitted signal power until the signal-to-noise power ratio is as high as desired. There is no doubt that this solution works, but it is often prohibitively expensive. Such developments as communications satellites and deep space probes have stimulated engineers to devise more and more ingenious ways of identifying weak signal patterns against a noisy background, whose power may exceed that of the signal by many orders of magnitude. Much of the fascination of these recent developments lies in the noise-combating techniques which are required.

Because noise is random in nature it is necessary, in order to formulate the problem properly, to make a short digression into the theory of statistics. This is done in Chapter 5 and followed by a brief account of the origins of electrical noise in Chapter 6.

This opens the way for a comparison of the behaviour of some common modulation systems towards noisy signals in Chapter 7. It turns out that certain modulation methods allow an improvement in signal-to-noise ratio at the expense of occupying more of the frequency spectrum. F.M. broadcast systems take advantage of this important effect.

A key problem, as noted above, is that of recognising a signal pattern which has become buried in noise. The general problem is a difficult one and the concluding sections of Chapter 7 deal with just one or two special cases in which the noise power is assumed to be uniformly distributed throughout the spectrum and the noise statistics are Gaussian. An optimum receiver can then be specified. We can be sure at the outset, however, that there is no signal pattern which cannot by chance be reproduced from time to time by the noise. A receiver designed to recognise certain patterns cannot distinguish between a 'real' message and an accidental noise copy of it. Most of the time it may announce the messages correctly—but how to avoid the occasional deception by a 'whisper in the wind'?

1.4 INFORMATION

A quite different approach to the communication process is taken in the concluding chapters, where attention is directed to the information which message-bearing signals carry, rather than to the signals themselves. The idea of a quantitative measure of information, and the concept of entropy in relation to information sources, lead to a discussion of coding and the achievement of maximum information transmission rates under noise-free conditions. By far the most surprising and valuable result of information theory, however, relates to communication over noisy channels, in that it can be shown that messages may be transmitted with negligible error in spite of the noise, even if the noise power greatly exceeds that of the signal. The noise necessitates a reduction in the rate of information transmission but coding procedures may be shown to exist which, in principle at least, make error-free message transmission possible. The 'whispers in the wind' no longer deceive.

Much of the introduction to information theory presented in Chapters 8 and 9 may be understood without reference to the earlier material. The reader may, if he wishes, begin with this part of the book, or else move into it by way of an interlude during the more earthbound study of signal theory which now follows.

Chapter 2

Describing Signals

2.1 INTRODUCTION

A signal waveform may be described by giving an analytical or graphical statement of the way the voltage or current varies with time. For many purposes, however, it is more helpful to regard the signal as though it were the result of adding together a number of simpler *component* waveforms. Probably the best-known example of this technique is the Fourier series expansion, in which the component waveforms are sine waves, and it will be shown below how any repetitive waveform may be broken down into a set of sinusoidal components. A display of the amplitudes and phases of the individual sinusoids along the frequency axis—i.e., the *spectrum* of the signal—then provides an alternative description, termed the *frequency domain* description of the signal, in contrast to the original *time domain* description.

The usefulness of spectral analysis in analysing the transmission of signals through linear networks will be demonstrated in Chapter 3. In later chapters, too, it will be found a useful thinking aid to be able to move from the time domain to the frequency domain or vice versa when analysing various types of signal. The purpose of the present chapter is to provide the necessary mathematical background, beginning with the Fourier series expansion and then developing this technique in order to deal with the spectral analysis of single pulses and of random waveforms.

2*

The emphasis here is on breaking down a signal into component sinusoids, but on the way we shall find there is an important class of signals (band-limited signals) which may alternatively be decribed as the linear superposition of elementary components each of which has the shape sin x/x. This result is embodied in the *sampling theorem* and has far-reaching consequences in pulse modulation theory.

2.2 FOURIER SERIES EXPANSION

Fourier's theorem states that a function f(x), defined in the range $-\pi < x < \pi$ may be expanded as a series of the form

$$f(x) = A_0 + \sum_{n=1}^{\infty} A_n \cos nx + \sum_{n=1}^{\infty} B_n \sin nx \qquad 2.1$$

A proof of the theorem is beyond the scope of this book, although it should be noted that there are certain conditions f(x) must satisfy for such an expansion to be valid. In particular the integral $\int |f(x)| \, dx$ over the whole interval must be finite and only a finite number of discontinuities in f(x) is permissible. These conditions however, will not exclude any communication signal of practical interest.

In cases where f(x) is discontinuous, the right-hand side of equation 2.1 at the point of discontinuity takes a value which is the average of the two values of f(x) obtained when the discontinuity is approached from either side. This behaviour will be seen in some of the examples below.

The method of evaluating the coefficients A_0, A_n and B_n is as follows. To determine, for example, the coefficient A_m, multiply both sides of equation 2.1 by cos mx and integrate over the range $-\pi < x < \pi$,

$$\int_{-\pi}^{\pi} f(x) \cos mx \, dx = \int_{-\pi}^{\pi} A_0 \cos mx \, dx$$

$$+ \sum_{n=1}^{\infty} A_n \int_{-\pi}^{\pi} \cos nx \cos mx \, dx$$

$$+ \sum_{n=1}^{\infty} B_n \int_{-\pi}^{\pi} \sin nx \cos mx \, dx$$

Each of the integrals on the right-hand side is zero except

$$A_m \int_{-\pi}^{\pi} \cos^2 mx \, \mathrm{d}x = \pi A_m$$

Hence we have

$$\int_{-\pi}^{\pi} f(x) \cos mx \, \mathrm{d}x = \pi A_m$$

and thus each of the coefficients A_m may be found, provided the integral on the left can be evaluated (either analytically or by some numerical method such as Simpson's rule).

Similarly it may be shown that

$$\int_{-\pi}^{\pi} f(x) \sin mx \, \mathrm{d}x = \pi B_m$$

while the remaining coefficient, A_0, comes from integrating both sides of equation 2.1 directly

$$\int_{-\pi}^{\pi} f(x) \, \mathrm{d}x = 2\pi A_0$$

The coefficient A_0 is seen to be the mean value of the function $f(x)$ within the range of expansion.

These results are summarised in the three equations

$$A_0 = \frac{1}{2\pi} \int_{-\pi}^{\pi} f(x) \, \mathrm{d}x$$

$$A_n = \frac{1}{\pi} \int_{-\pi}^{\pi} f(x) \cos nx \, \mathrm{d}x \qquad 2.2$$

$$B_n = \frac{1}{\pi} \int_{-\pi}^{\pi} f(x) \sin nx \, \mathrm{d}x$$

Only a few terms are often sufficient for the series to give an acceptable approximation to the function $f(x)$. Consider, for example, the function

$$f(x) = \frac{x}{\pi} \qquad -\pi < x < \pi$$

11

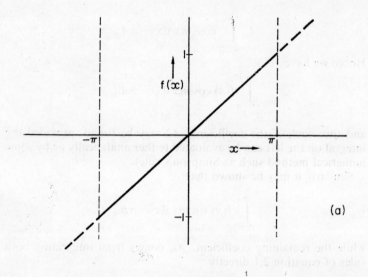

(a)

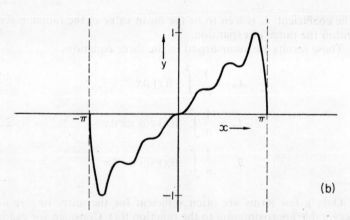

(b)

Fig. 2.1. The function f(x) = x/π *approximated within the range* −π < x < π
by the first six terms of a Fourier series

shown in Fig. 2.1(a). Inserting this function into equations 2.2 we find

$$A_0 = A_1 = A_2 = A_3 = \ldots = 0$$

$$B_1 = \frac{2}{\pi}$$

$$B_2 = -\frac{1}{\pi}$$

$$B_3 = \frac{2}{3\pi}$$

$$\ldots$$

$$B_n = (-1)^{n+1}\frac{2}{n\pi}$$

Hence equation 2.1 gives

$$f(x) = \frac{2}{\pi}\left(\sin x - \frac{\sin 2x}{2} + \frac{\sin 3x}{3} - \ldots\right)$$

The sum, y, of the first six terms of this series is shown in Fig. 2.1(b).

2.3 FOURIER ANALYSIS OF REPETITIVE WAVEFORMS

In equation 2.1 the function $f(x)$ was left undefined outside the interval $-\pi < x < \pi$. If we now allow x to take values outside this range then it is clear that the right-hand side will repeat periodically, taking in each interval $k\pi < x < (k+2)\pi$, where k is any odd integer, precisely the same sequence of values as in the fundamental interval, $-\pi < x < \pi$. The Fourier series expansion can therefore give a complete description of a periodic function.

Consider the voltage (or current) waveform $v(t)$ of Fig. 2.2(a) which is repetitive with period T. The choice of time origin is arbitrary. If the variable x is defined as $x = 2\pi t/T = \omega_1 t$, where $\omega_1 = 2\pi/T$, then, while t ranges from $-T/2$ to $T/2$, x ranges from $-\pi$ to π. We may therefore substitute $x = \omega_1 t$ into equations 2.1 and 2.2 and obtain

$$v(t) = A_0 + \sum_{n=1}^{\infty} A_n \cos n\omega_1 t + \sum_{n=1}^{\infty} B_n \sin n\omega_1 t \qquad 2.3$$

13

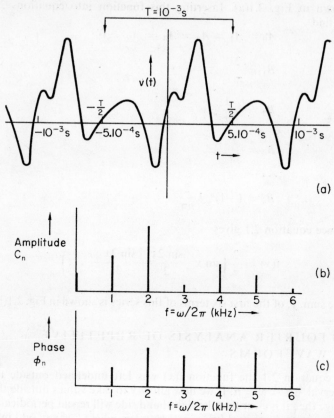

Fig. 2.2. (a) *Periodic signal* $v(t)$ *represented in the time domain;* (b) *and* (c) *The same signal represented in the frequency domain*

where

$$A_0 = \frac{1}{T} \int_{-T/2}^{T/2} v(t) \, dt$$

$$A_n = \frac{2}{T} \int_{-T/2}^{T/2} v(t) \cos n\omega_1 t \, dt \qquad 2.4$$

$$B_n = \frac{2}{T} \int_{-T/2}^{T/2} v(t) \sin n\omega_1 t \, dt$$

14

Equation 2.3 may also be written in the alternative form

$$v(t) = A_0 + \sum_{n=1}^{\infty} C_n \cos{(n\omega_1 t - \phi_n)} \qquad 2.5$$

where
$$C_n = \sqrt{(A_n^2 + B_n^2)}$$

and
$$\tan{\phi_n} = \frac{B_n}{A_n}$$

The repetitive waveform $v(t)$ has now been broken down into a series of component sinusoids whose frequencies are multiples of ω_1 together with the constant or d.c. term A_0. The component having the lowest frequency, $C_1 \cos{(\omega_1 t - \phi_1)}$, is termed the *fundamental*, the other components are *harmonics*, that at $n\omega_1$ being termed the nth harmonic. A knowledge of the amplitudes of the harmonics along the frequency axis, Fig. 2.2(b), together with a knowledge of the phases ϕ_n, Fig. 2.2(c), constitutes a description of the signal in the frequency domain. Note that this is just as complete a description as that in the time domain, Fig. 2.2(a). Either description contains all the information about the signal. A cathode ray oscilloscope will display a signal in the time domain after the manner of Fig. 2.2(a). The wave analyser, on the other hand, is an instrument designed for analysis in the frequency domain. This instrument acts effectively as a band-pass filter of fixed, narrow, bandwidth, whose mid-frequency may be varied to measure the amplitudes of the harmonics of an incoming signal. Sometimes the frequency range of interest is scanned automatically to give a display like that of Fig. 2.2(b). A wave analyser takes no account of phase, however, so that it could not distinguish between the signals of Figs 2.3(a) and (b), which have the same harmonic amplitudes, but different phases. For some purposes this is of no great consequence; the human ear, for example, is insensitive to phase relationships and cannot distinguish between the signals of Fig. 2.3.

In the case of waveforms which possess odd or even symmetry about some moment in time, a useful simplification is possible by choosing this moment as the time origin. If the waveform is odd, the cosine coefficients A_n are then all zero since the integrand $v(t) \cos{n\omega_1 t}$ in the second of equations 2.4 is odd and this integral vanishes. Conversely, if the waveform is even, the sine coefficients B_n are all zero. An example of odd symmetry was seen in Fig. 2.3(b),

15

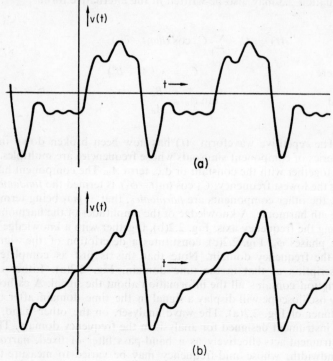

(a)

(b)

Fig. 2.3. Two periodic signals with identical harmonic amplitudes but different phases

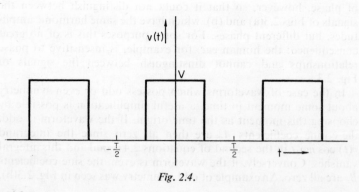

Fig. 2.4.

which is a sawtooth wave whose Fourier series expansion was derived in Section 2.2, namely

$$v(t) = \frac{2V}{\pi} \left(\sin \omega_1 t - \frac{\sin 2\omega_1 t}{2} + \frac{\sin 3\omega_1 t}{3} - \ldots \right) \qquad 2.6$$

Here the cosine coefficients, as expected, are all zero. An example of even symmetry is found in the square wave of Fig. 2.4. The B_n coefficients are zero in this case, while for the A_n coefficients

$$A_0 = \frac{V}{2}$$

$$A_n = \frac{2}{T} \int_{-T/2}^{T/2} v(t) \cos n\omega_1 t = \frac{2V}{T} \int_{-T/4}^{T/4} \cos n\omega_1 t \, dt$$

$$= \frac{2V}{n\pi} \sin n\pi/2$$

Hence, for the square wave

$$v(t) = \frac{V}{2} + \frac{2V}{\pi} \left(\cos \omega_1 t - \frac{1}{3} \cos 3\omega_1 t + \frac{1}{5} \cos 5\omega_1 t \ldots \right) \qquad 2.7$$

It is of interest to plot the function given by the right-hand side of equation 2.7 when only a finite number of terms is included. Fig. 2.5 shows the result of including terms up to $n = 7, 15$, and 31, omitting the d.c. term A_0. The approximation to a square wave clearly improves as the number of terms is increased except that, close to the discontinuity, the curve always overshoots. The size of the overshoot, about 9% of the voltage step V, does not diminish as the number of terms is increased, although it becomes steadily narrower. This means that the *energy* content of this 'spike' ultimately vanishes and therefore ceases to be of physical significance. The behaviour of a truncated Fourier series close to a discontinuity is known as 'Gibbs's phenomenon' after J. Willard Gibbs.[1] Fig. 2.3(b) shows the same effect with a sawtooth waveform.

The overshoots in Fig. 2.5 are the result of adding together only the first few terms of equation 2.7. The same phenomenon would appear if we were to pass a square wave through a linear network which attenuates strongly above a certain frequency but passes freely the components below that frequency. In other words, a square wave—equation 2.7 with an infinite number of terms—is applied to a low-pass filter. The output signal would be given by the series of equation 2.7 truncated at the filter cut-off frequency and would therefore exhibit overshoot and 'ringing' at each disconti-

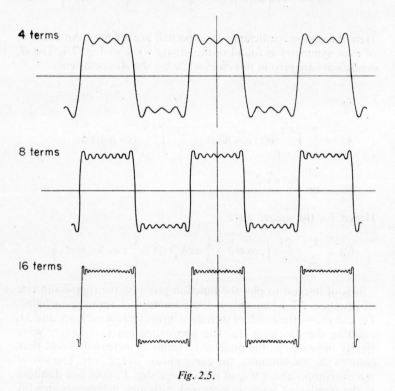

Fig. 2.5.

nuity, as shown in Fig. 2.5. In practice some further distortion also results due to phase changes in the filter network, making the output rather less symmetrical. Signal transmission will be dealt with further in the following chapter, but this example suffices to show that Gibbs's phenomenon is of more than academic interest.

2.4 IMPULSES, IMPULSE TRAINS, AND SOME FURTHER FOURIER SERIES

Fig. 2.6(a) shows a rectangular pulse of height V and duration τ, centred on the moment $t = t_0$. Let the area under the pulse be denoted by k, thus

$$\int_{-\infty}^{\infty} v(t)\,\mathrm{d}t = V\tau = k$$

Now let V be increased to V' and τ be simultaneously decreased to τ' as in Fig. 2.6(b), such that $V\tau = V'\tau' = k$ is kept constant.

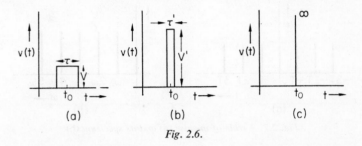

Fig. 2.6.

The limit of this process, as $V \to \infty$ and $\tau \to 0$, keeping the area constant, yields the impulse signal represented in Fig. 2.6(c).

An electrical impulse may be thought of as a signal of negligible duration but of sufficiently large amplitude for its integral $\int v(t)\,\mathrm{d}t$ to be finite. This integral we shall call the *strength* of the impulse, with dimensions volt-second (or ampere-second). A convenient notation for an impulse of strength k occurring at $t = t_0$ is $k\delta(t-t_0)$. The unit impulse or 'δ-function', $\delta(t-t_0)$, is defined to be zero except within a region indefinitely close to $t = t_0$. Thus,

$$\delta(t-t_0) = 0 \quad \text{if} \quad t \neq t_0$$

while
$$\int_{t_1}^{t_2} \delta(t-t_0)\,\mathrm{d}t = 1 \quad \text{if} \quad t_1 < t_0 < t_2$$

It follows that if $\mathrm{f}(t)$ is some abitrary time function then

$$\int_{t_1}^{t_2} \mathrm{f}(t)\,\delta(t-t_0)\,\mathrm{d}t = \mathrm{f}(t_0) \qquad 2.8$$

provided the range of integration embraces t_0.

19

Signals and Information

Although the impulse function has been introduced as an extreme form of rectangular pulse, the actual shape need be of no concern and several representations are possible. The assumption of negligible duration implies an impulse too fast for the observing system to follow. That is to say, the shape of our impulse signal cannot be observed and is of no consequence. Only the strength of the impulse is important. It is the electrical analogue of the impulse or hammer-blow in mechanics, where the applied force F is indefinitely large for an indefinitely short time and the only significant parameter is the integral $\int F\, dt$ which represents the momentum imparted by the blow.

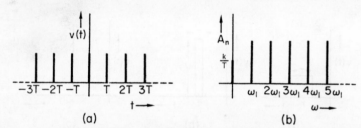

Fig. 2.7. A train of impulse (a) and its spectrum (b)

Fig. 2.7(a) shows a train of impulses each of strength k, occurring at $t = 0,\ \pm T,\ \pm 2T$, etc. Applying equations 2.4 and using 2.8 we have

$$A_0 = \frac{k}{T}$$

$$A_n = \frac{2k}{T} \int_{-T/2}^{T/2} \delta(t) \cos n\omega_1 t\, dt$$

$$= \frac{2k}{T}$$

$$B_n = 0$$

The Fourier expansion is thus

$$v(t) = \frac{2k}{T} \left(\frac{1}{2} + \cos \omega_1 t + \cos 2\omega_1 t + \cos 3\omega_1 t + \ldots \right) \qquad 2.9$$

in which all the harmonics have the same amplitude (Fig. 2.7(b)).

20

This result may be used to derive, in a relatively easy manner, the Fourier series representing a train of rectangular pulses. First consider the signal of Fig. 2.8(a) which comprises two impulse trains of equal strength and opposite polarity, displaced in time

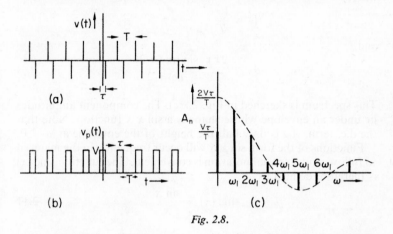

Fig. 2.8.

by τ. The Fourier series for this signal may be written down immediately from equation 2.9

$$v(t) = \frac{2k}{T} \sum_n [\cos n\omega_1(t+\tau/2) - \cos n\omega_1(t-\tau/2)]$$

$$= -\frac{4k}{T} \sum_n \sin n\omega_1\tau/2 \; \sin n\omega_1 t$$

Integrating this function, Fig. 2.8(a), produces the rectangular pulse train of Fig. 2.8(b). The Fourier series for this signal may therefore be derived by integrating term by term the series just obtained, thus

$$v_p(t) = \frac{4k}{T} \sum_n \frac{\sin n\omega_1\tau/2}{n\omega_1} \; \cos n\omega_1 t + \text{const.}$$

This gives the Fourier expansion of a train of rectangular pulses of amplitude k, duration τ, and repetition rate $\omega_1/2\pi = 1/T$. The constant of integration is chosen to correspond to the coefficient A_0 which clearly has the value $k\tau/T$. For later reference the form of

21

this expansion is written below, replacing the pulse amplitude k by V,

$$v(t) = A_0 + \sum_n A_n \cos n\omega_1 t \qquad 2.10$$

where

$$A_0 = \frac{V\tau}{T}$$

and

$$A_n = \frac{2V\tau}{T} \frac{\sin(n\omega_1\tau/2)}{n\omega_1\tau/2}$$

This spectrum is sketched in Fig. 2.8(c). The component amplitudes lie under an envelope whose shape is a sin x/x function. Note that the d.c. term, A_0, is only half the height of the envelope at $\omega = 0$.

Functions of the form sin x/x will occur frequently in the material to follow, and a special notation is convenient. The function sinc (x) is therefore defined as

$$\text{sinc}(x) = \frac{\sin x}{x} \qquad 2.11$$

A table of values of this function, sketched in Fig. 2.18(a), is given in Appendix 1, while some of its more important properties will be discussed in Section 2.16. The expression above for A_n may now be written

$$A_n = \frac{2V\tau}{T} \text{ sinc}(n\omega_1\tau/2) \qquad 2.12$$

When the pulse duration, τ, is made equal to $T/2$ the pulse train becomes a square wave and it is easily shown that equation 2.10 then becomes identical to equation 2.7. Removing the d.c. level and adjusting the amplitude so that the signal swings from $+1$ to -1, the expansion becomes that shown in Fig. 2.9(a). Further integration gives the triangular waveform of Fig. 2.9(b) where the d.c. level and the amplitude have again been adjusted to give a total excursion of ±1. Repeating the process once more, the waveform of Fig. 2.9(c) is obtained, which consists of a sequence of smoothly joined parabolic hoops. Examination of the Fourier series given in Fig. 2.9 shows that, as the discontinuities are removed and the waveform is smoothed, the amplitudes of the higher harmonics are considerably reduced. Any smoothly varying signal will be relatively lacking in high-frequency components, while one which exhibits

22

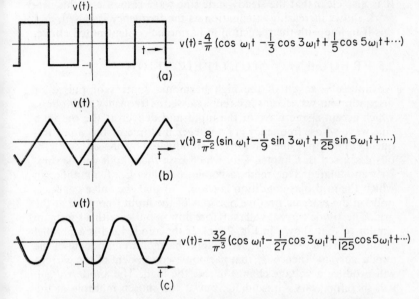

$$v(t) = \frac{4}{\pi}\left(\cos \omega_1 t - \frac{1}{3}\cos 3\omega_1 t + \frac{1}{5}\cos 5\omega_1 t + \cdots\right)$$

(a)

$$v(t) = \frac{8}{\pi^2}\left(\sin \omega_1 t - \frac{1}{9}\sin 3\omega_1 t + \frac{1}{25}\sin 5\omega_1 t + \cdots\right)$$

(b)

$$v(t) = -\frac{32}{\pi^3}\left(\cos \omega_1 t - \frac{1}{27}\cos 3\omega_1 t + \frac{1}{125}\cos 5\omega_1 t + \cdots\right)$$

(c)

Fig. 2.9. Square wave and related waveforms by successive integration

sudden jumps, or discontinuities in its first derivative, will be rich in harmonic content, for such a waveform can only be synthesised if the rapidly varying high-frequency components are included.

The process of integration described above may be performed electrically. It may be shown that, over times which are short compared with the time constant RC, the output from the network of Fig. 2.10 is proportional to the integral of the input signal. It follows that if the input is the square wave of Fig. 2.9(a) the output will have the triangular form of Fig. 2.9(b), provided $RC \gg T$.

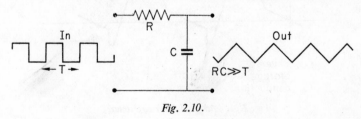

Fig. 2.10.

It is also clear that the steady-state sine wave response of this network shows increasing attenuation as the frequency is raised, and this is in line with the reduction in harmonic content noted above.

2.5 FREQUENCY MULTIPLICATION

Advantage is taken of the high harmonic content of pulsed or discontinuous waveforms in circuits classed as frequency multipliers, which accept a sine wave at the input and deliver at the output a sine wave whose frequency is an integer multiple of the input frequency. The input sine wave may be modulated by one of the methods discussed in Chapter 4, in which case the output will be similarly modulated. One method employs a class C r.f. amplifier in which the grid bias condition is chosen so that the valve conducts only on the extreme positive portions of the input sine wave at the grid. The anode current will therefore flow in pulses with a spectrum similar to that shown in Fig. 2.8(c). If the anode load is a parallel resonant circuit, which is tuned to some harmonic component of the anode current, then only that harmonic component of the current will produce a voltage change across the load. The anode *voltage* will therefore vary sinusoidally, but at the chosen multiple of the driving frequency.

Frequency multiplication may also be used to produce signals at microwave frequencies from a source of much lower frequency. Special semiconductor diodes, known as step recovery diodes, have been developed, which exhibit a very rapid transition from reverse storage conduction to cut-off. Fig. 2.11 shows a typical current

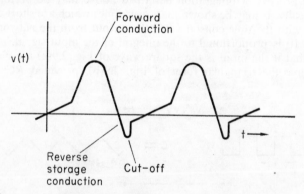

Fig. 2.11. (Adapted from Krakauer[2])

waveform in a circuit containing a step recovery diode and a sinusoidal driving e.m.f. The step transition which occurs just prior to cut-off occupies less than a nanosecond and the waveform is rich in harmonic content. With excitation at 200 MHz, for example, the tenth harmonic at 2 GHz may be selected and with a suitably matched load (here a microwave cavity) as much as 10% of the input power appears at the output frequency. Further details may be found in reference 2 at the end of this chapter.

2.6 TIME DELAY

In discussing Fig. 2.3 it was pointed out that a disturbance of the relative phases of the Fourier components will usually result in distortion of the wave shape. There is, however, one important condition which results in a shift of the wave along the time axis, without any change of shape. This is found by examining the expansion given by equation 2.5

$$v(t) = A_0 + \sum_n C_n \cos (n\omega_1 t - \phi_n) \qquad 2.5$$

from which the expansion of the same waveform delayed by t_d may be written as

$$v(t - t_d) = A_0 + \sum_n C_n \cos [n\omega_1(t - t_d) - \phi_n]$$

$$= A_0 + \sum_n C_n \cos (n\omega_1 t - \phi_n + \psi_n) \qquad 2.13$$

where $\psi_n = -n\omega_1 t_d$.

It follows that introducing a phase lag into each Fourier component in equation 2.5, where the phase lag is proportional to the frequency of the component, results in delaying the signal in time. The time domain operation, 'shift the waveform along the time axis', is therefore equivalent to the frequency domain operation 'introduce a phase change proportional to frequency'. With frequency measured in radians per second, the constant of proportionality is equal to the time delay in seconds.

2.7 ORTHOGONAL SIGNALS

Two functions $v(t)$ and $w(t)$ are said to be *orthogonal* in the interval $t_1 < t < t_2$ if

$$\int_{t_1}^{t_2} v(t)\, w(t)\, dt = 0 \qquad 2.14$$

3*

Any pair of the sinusoidal components of a repetitive waveform are orthogonal in the fundamental interval, $-T/2 < t < T/2$, or in any whole number of intervals T. For example

$$\int_{-T/2}^{T/2} \cos k\omega_1 t \sin l\omega_1 t\,\mathrm{d}t = 0$$

Other examples of orthogonal signals are not hard to find; a pair are shown in Fig. 2.12, and others will appear in the material to

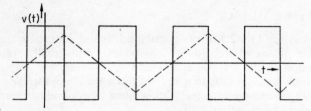

Fig. 2.12. Orthogonal signals

follow. The use of the geometrical term 'orthogonal' comes from a method of representing signals as vectors in a many-dimensional hyperspace, which is presented in Appendix 3.

2.8 MEAN POWER

If a periodic voltage $v(t)$ is applied to a resistor R the average power P_R dissipated in R will be

$$P_R = \frac{1}{RT} \int_{-T/2}^{T/2} v^2(t)\,\mathrm{d}t$$

It is convenient to omit the factor $1/R$ and refer to the mean power P as the mean square value of the signal. Thus, if $v(t)$ is a voltage (or current) waveform then the dimensions of P are volt^2 (or amp^2) while P is numerically equal (in either case) to the power which would be delivered by the signal to a $1\ \Omega$ resistor.

The mean power of a periodic signal may be expressed in terms of the Fourier coefficients as follows

$$P = \frac{1}{T} \int_{-T/2}^{T/2} v^2(t)\,\mathrm{d}t$$

$$= \frac{1}{T} \int_{-T/2}^{T/2} \left[A_0 + \sum_n A_n \cos n\omega_1 t + \sum_n B_n \sin n\omega_1 t \right]^2 \mathrm{d}t$$

On expanding the bracket, the integrals involving the cross-product terms are all zero, since the Fourier components are mutually orthogonal. The remaining terms yield

$$P = A_0^2 + \sum_n \frac{A_n^2 + B_n^2}{2}$$

$$= A_0^2 + \frac{1}{2} \sum_n C_n^2 \qquad\qquad 2.15$$

in which the second equation uses the coefficients C_n of equation 2.5.

Equation 2.15 shows that the mean power of a periodic signal is equal to the sum of the powers of the separate Fourier components. Note that this is only true because the component waveforms are orthogonal. If two arbitrary signals are added together it will *not* in general be true that the total power is the sum of the powers of the individual signals.

2.9 COMPLEX FOURIER COEFFICIENTS. 'NEGATIVE FREQUENCY'

Each of the terms $A_n \cos n\omega_1 t$ in equation 2.3 may be replaced by

$$A_n \cos n\omega_1 t = \frac{A_n}{2} e^{jn\omega_1 t} + \frac{A_n}{2} e^{-jn\omega_1 t}$$

That is to say, we consider $A_n \cos n\omega_1 t$ as the sum of two phasors in the complex plane, rotating in opposite directions with angular velocities $\pm n\omega_1$, each of length $A_n/2$. Similarly the terms $B_n \sin n\omega_1 t$ may be written

$$B_n \sin n\omega_1 t = -\frac{jB_n}{2} e^{jn\omega_1 t} + \frac{jB_n}{2} e^{-jn\omega_1 t}$$

Equation 2.3 now becomes

$$v(t) = A_0 + \sum_{n=1}^{\infty} \frac{A_n - jB_n}{2} e^{jn\omega_1 t} + \sum_{n=1}^{\infty} \frac{A_n + jB_n}{2} e^{-jn\omega_1 t}$$

Here the integer n is restricted to positive values, but alternatively we may write

$$v(t) = \sum_{k=-\infty}^{\infty} \alpha_k e^{jk\omega_1 t} \qquad\qquad 2.16$$

where the summation extends over all integers k, positive and

27

negative, and the complex coefficients α_k are such that when k is positive, say $k = n$,

$$\alpha_n = \frac{A_n - jB_n}{2} \qquad\qquad 2.17$$

when k is negative, say $k = -n$,

$$\alpha_{-n} = \frac{A_n + jB_n}{2} = \alpha_n^* \qquad\qquad 2.18$$

while, for $k = 0$

$$\alpha_0 = A_0$$

The signal $v(t)$ is here regarded as the sum of many rotating phasors. The sum is always real, for consider the sum of the pair of phasors

$$\begin{aligned}
\alpha_n e^{jn\omega_1 t} + \alpha_{-n} e^{-jn\omega_1 t} &= \alpha_n e^{jn\omega_1 t} + (\alpha_n e^{jn\omega_1 t})^* \\
&= 2\mathcal{R}[\alpha_n e^{jn\omega_1 t}] \\
&= \mathcal{R}[(A_n - jB_n)e^{jn\omega_1 t}] \\
&= C_n \cos(n\omega_1 t - \phi_n)
\end{aligned}$$

The usual notations for taking the complex conjugate and the real part have been used here, while the last step refers to equation 2.5. It follows that each component in equation 2.16, when taken together with its partner rotating in the opposite sense at the same rate, gives the corresponding term in equation 2.5.

Multiplying both sides of equation 2.16 by $e^{-jk'\omega_1 t}$ and integrating from $-T/2$ to $T/2$ gives an expression for any particular coefficient $\alpha_{k'}$,

$$\begin{aligned}
\int_{-T/2}^{T/2} v(t)e^{-jk'\omega_1 t} &= \sum_{k=-\infty}^{\infty} \int_{-T/2}^{T/2} \alpha_k e^{j(k-k')\omega_1 t}\, dt \\
&= \int_{-T/2}^{T/2} \alpha_{k'}\, dt \\
&= \alpha_{k'} T
\end{aligned}$$

Hence

$$\alpha_{k'} = \frac{1}{T} \int_{-T/2}^{T/2} v(t)e^{-jk'\omega_1 t}\, dt$$

This result may also be obtained by substituting A_n and B_n from equations 2.4 into 2.17 and 2.18. The integer k' in this last equation may be positive, negative, or zero.

28

Summarising these results, we have shown that *a repetitive waveform* $v(t)$ *with period T may be expanded as a complex Fourier series in* $e^{jn\omega_1 t}$, *where* $\omega_1 = 2\pi/T$, *of the form*

$$v(t) = \sum_{n=-\infty}^{\infty} \alpha_n e^{jn\omega_1 t} \qquad \qquad 2.19$$

where $$\alpha_n = \frac{1}{T} \int_{-T/2}^{T/2} v(t) e^{-jn\omega_1 t} \qquad \qquad 2.20$$

In general, the α-coefficients are complex numbers and contain information about the magnitude and phase of the components in equation 2.5, namely

$$\phi_n = -\arg \alpha_n \qquad (n > 0) \qquad \qquad 2.21$$
$$C_n = 2 |\alpha_n| \qquad \qquad 2.22$$

In section 2.3, however, it was shown that when the function $v(t)$ is even, $B_n = 0$ and $\phi_n = 0$ for all n and thus equations 2.17 and 2.18 show that in this case the α-coefficients are all real numbers. Conversely, when $v(t)$ is odd, the α-coefficients are purely imaginary. These results also follow from substituting $e^{-jn\omega_1 t} = \cos n\omega_1 t - j \sin n\omega_1 t$ in equation 2.20 and examining the symmetry of the two resulting integrands. It is also useful to note that, setting $t = 0$ in equation 2.17,

$$v(0) = \sum_n \alpha_n \qquad \qquad 2.23$$

As an illustration of the complex Fourier expansion, consider once more the train of rectangular pulses of Fig. 2.8(b). The α-coefficients are found using equation 2.18.

$$\begin{aligned}
\alpha_n &= \frac{1}{T} \int_{-T/2}^{T/2} v(t) e^{-jn\omega_1 t} \, dt \\
&= \frac{V}{T} \int_{-\tau/2}^{\tau/2} e^{-jn\omega_1 t} \, dt \\
&= \frac{2V}{T} \frac{\sin n\omega_1 \tau/2}{n\omega_1} \qquad \qquad 2.24
\end{aligned}$$

or (cf. equation 2.12)

$$\alpha_n = \frac{V\tau}{T} \operatorname{sinc}(n\omega_1\tau/2)$$

This spectrum is sketched in Fig. 2.13(a) for the case $T = 5\tau$. The choice of time origin has made $v(t)$ an even function so that the α-coefficients in this example are purely real and are represented by vertical lines at the appropriate positions along the axis of ω. This axis now includes negative as well as positive values of ω, since our expansion involves phasors $\alpha e^{j\omega t}$ each of which has a partner rotating in the opposite direction (but of the same magnitude —hence the symmetry of the spectrum about $\omega = 0$). The ω-axis is best labelled 'phasor angular velocity'. It is more convenient, however, to retain the term 'angular frequency' or, scaling by a factor 2π, such a spectrum may be plotted against frequency in Hz.

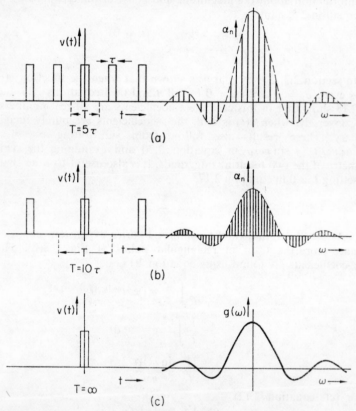

Fig. 2.13. Trains of rectangular pulses, of duration τ and repetition period T, with their spectra. (a) $T = 5\tau$, (b) $T = 10\tau$, and (c) $T = \infty$

This is simply for convenience, however, and does not give any physical reality to 'negative frequencies.'

The mean power P of a periodic signal may also be expressed in terms of the complex coefficients. Using equations 2.15 and 2.22 we have

$$P = A_0^2 + \frac{1}{2} \sum_{n=1}^{\infty} C_n^2 = \alpha_0^2 + 2 \sum_{n=1}^{\infty} |\alpha_n|^2$$

$$= \sum_{n=-\infty}^{\infty} |\alpha_n|^2$$

Thus
$$P = \frac{1}{T} \int_{-T/2}^{T/2} v^2(t)\,\mathrm{d}t = \sum_n |\alpha_n|^2 \qquad 2.25$$

which is one form of a more general result known as Parseval's theorem (see also Section 2.12).

2.10 FOURIER TRANSFORM

Fig. 2.13(a) shows the complex Fourier spectrum, calculated above, for a train of rectangular pulses of amplitude V and duration τ, repeating at intervals $T = 5\tau$ and with one of the pulses centred on the origin. The complex coefficients are given by equation 2.24 and components appear at angular frequencies $n\omega_1$ where $\omega_1 = 2\pi/T$.

This example provides an opportunity to investigate the behaviour of the spectrum of a train of pulses as the pulse interval T is made indefinitely large, and thus to develop a technique for representing a single pulse—rather than a repetitive waveform—in the frequency domain. Fig. 2.13(b) shows the spectrum of a train of pulses of the same pulse duration τ but repeating at intervals $T = 10\tau$. Two changes have occurred: (i) ω_1 is now smaller, so the lines have moved closer together, and (ii) each component is reduced in size because of the factor τ/T in equation 2.24. This process may be continued, increasing T still further, until a stage is reached where the discrete Fourier component description ceases to be useful. If the spectrum were being examined by a wave analyser, the individual components would become either too small to be observable, or too close together to be resolved. The *sum* of the coefficients, however, remains constant (equation 2.23).

It is rather as though we are watching a string of beads, when a demon comes along and replaces each bead by two smaller beads, while keeping the total mass of the string constant. If the demon

returns often enough, a stage is reached when it it no longer convenient to think in terms of a string of beads but rather of a continuous distribution of mass—indeed this is how we normally regard
strings or wires, even though we know them to be made up of
atoms. Having adopted the continuous-distribution-of-mass idea,
we no longer enquire about the masses of the separate beads, but
about the *density* or mass per unit length of the string. We may
then forget about the demon, for no matter how often he returns,
the density of the string does not change—though it may vary
along the length of the string, according to the shape of the original
necklace.

Applying this kind of argument to the spectra of Figs 2.13(a)
and (b), we evaluate the *spectral density* at angular frequency
$\omega = n\omega_1$ by dividing α_n by the interval ω_1 thus

$$\text{spectral density} = g(\omega) = \frac{\alpha_n}{\omega_1} = \frac{V\tau}{\omega_1 T} \text{ sinc } (n\omega_1\tau/2)$$

$$= \frac{V\tau}{2\pi} \text{ sinc } (n\omega_1\tau/2)$$

This will be approximate for finite ω_1, becoming exact as $\omega_1 \rightarrow 0$
($T \rightarrow \infty$). In taking this limit we must also allow n to become
indefinitely large such that $n\omega_1 = \omega$ is held constant. The limiting
process leaves, in the time domain, a single rectangular pulse
centred on the origin. The corresponding spectral density in the
frequency domain is given by

$$g(\omega) = \frac{V\tau}{2\pi} \text{ sinc } (\omega\tau/2) \qquad 2.26$$

This continuous function of ω is shown in Fig. 2.13(c).

The spectral density of a single pulse, or transient signal $v(t)$ of
any shape, is found in a similar way, dividing the expression for
α_n given by equation 2.20, by ω_1. Recalling that $\omega_1 = 2\pi/T$ and
allowing $\omega_1 \rightarrow 0$ we have,

$$g(\omega) = \lim_{\omega_1 \rightarrow 0} \frac{\alpha(\omega)}{\omega_1} = \frac{1}{2\pi} \int_{-\infty}^{\infty} v(t) \, e^{-j\omega t} \, dt \qquad 2.27$$

in which the limits of integration have become infinite. The form of
equation 2.19 also changes. Writing

$$\alpha_n = \alpha(\omega) = \omega_1 g(\omega)$$

equation 2.19 becomes

$$v(t) = \sum g(\omega) e^{j\omega t}\omega_1$$

where the summation is over the discrete values of $\omega = n\omega_1$. As $\omega_1 \to 0$, the sum may be replaced by the integral

$$v(t) = \int_{-\infty}^{\infty} g(\omega) e^{j\omega t} d\omega \qquad 2.28$$

Equations 2.27 and 2.28 are properly called integral transforms; in particular, $g(\omega)$ is called the *Fourier transform* (or Fourier integral) of $v(t)$, while $v(t)$ is the *inverse Fourier transform of* $g(\omega)$. The Fourier transform is closely related to the Laplace transform which finds such powerful applications in electrical circuit theory and the reader is referred to texts devoted to circuit analysis for discussion on this point.

The definition of the Fourier transform most commonly adopted in communication theory differs slightly from that of equation 2.27 by the omission of the factor $1/2\pi$. In this case a factor $1/2\pi$ must be included in equation 2.28. For the remainder of this text we define the Fourier transform $V(\omega)$ of a signal $v(t)$ as $V(\omega) = 2\pi g(\omega)$, thus

$$V(\omega) = \int_{-\infty}^{\infty} v(t) e^{-j\omega t} dt \qquad 2.29$$

while

$$v(t) = \frac{1}{2\pi} \int_{-\infty}^{\infty} V(\omega) e^{j\omega t} d\omega \qquad 2.30$$

It is also useful to follow the convention of using a capital letter to denote the transform, $X(\omega)$, of the signal denoted by the corresponding small letter, $x(t)$.

Any application of the Fourier transform method will be valid only if the function $v(t)$ is such that the integral in equation 2.29 exists. If $v(t)$ is zero everywhere except within some finite time interval and is finite within that interval, then the Fourier integral will converge for all values of ω. The Fourier transform will therefore provide a valid frequency domain description for a pulse or transient waveform of any shape likely to be of practical interest. It may be shown that a sufficient condition for $V(\omega)$ to exist is that the integral

$$\int_{-T}^{T} |v(t)| dt$$

remains finite as $T \to \infty$. Signals of infinite duration are therefore likely to be excluded, although certain waveforms not satisfying this condition can be handled satisfactorily by employing some suitable limiting process (for example, a step function and a continuous sine wave may be treated in this manner—see Problem 2.6 and Section 2.12 below).

2.11 SOME PROPERTIES OF THE FOURIER TRANSFORM

Before proceeding to calculate the spectral density distributions for some simple pulse shapes, it is worthwhile to examine some of the properties of the integral transforms expressed by equations 2.29 and 2.30. The Fourier transform was introduced in the previous section as a 'density of α-coefficients' so that many of its properties are shared in common with the complex Fourier coefficients and have been mentioned earlier.

(i) If the complex conjugate is taken on both sides of equation 2.29 then, remembering that $v(t)$ is a physical quantity (volts or amperes) and thus purely real, it follows that changing the sign of j is equivalent to changing the sign of ω, hence

$$V(-\omega) = V^*(\omega) \qquad 2.31$$

The corresponding result for the complex Fourier coefficients was noted in equation 2.18.

(ii) If $v(t)$ has even symmetry, then $V(\omega)$ is real and even. This follows by writing $e^{-j\omega t} = \cos \omega t - j \sin \omega t$ in equation 2.29. The imaginary part then involves the product of an even function, $v(t)$, and an odd function, $\sin \omega t$. Their product is odd and so this part of the integral vanishes. The Fourier transform $V(\omega)$ is in this case real and so by virtue of equation 2.31 it is also even.

Conversely, if $v(t)$ is odd, $V(\omega)$ is imaginary and odd. The corresponding result for the α-coefficients was given in the discussion following equation 2.22.

(iii) If an even function $x(t)$ has Fourier transform $X(\omega)$ then the function $X(t)$ has Fourier transform $2\pi x(\omega)$. The proof begins with equation 2.30 which in this case reads

$$x(t) = \frac{1}{2\pi} \int_{-\infty}^{\infty} X(\omega)\, e^{j\omega t}\, d\omega$$

substituting $u = -\omega$,

$$x(t) = -\frac{1}{2\pi} \int_{\infty}^{-\infty} X(-u) e^{-jut} \, du$$

$$= \frac{1}{2\pi} \int_{-\infty}^{\infty} X(-u) e^{-jut} \, du$$

but since the function x is real and even, it follows from property (ii) above that X is also real and even, hence

$$x(t) = \frac{1}{2\pi} \int_{-\infty}^{\infty} X(u) e^{-jut} \, du \qquad 2.32$$

Now consider X as a time function $X(t)$ with Fourier transform $Z(\omega)$, thus (equation 2.29)

$$Z(\omega) = \int_{-\infty}^{\infty} X(t) e^{-j\omega t} \, dt$$

$$= \int_{-\infty}^{\infty} X(u) e^{-j\omega u} \, du$$

where the last step simply renames the variable t as u. Finally, if t is replaced by ω in equation 2.32, we have

$$x(\omega) = \frac{1}{2\pi} \int_{-\infty}^{\infty} X(u) e^{-ju\omega} \, du = \frac{1}{2\pi} Z(\omega)$$

Hence

$$Z(\omega) = 2\pi x(\omega)$$

which is the desired result. For even functions therefore, two transforms are obtained for the price of one—a property which will be found useful in Section 2.12 below.

(iv) Setting $\omega = 0$ in equation 2.29 and $t = 0$ in 2.30 provides the particular relationships

$$V(0) = \int_{-\infty}^{\infty} v(t) \, dt \qquad 2.33$$

$$v(0) = \frac{1}{2\pi} \int_{-\infty}^{\infty} V(\omega) \, d\omega \qquad 2.34$$

(cf. equation 2.23)

35

(v) The signal $v(t) = v_1(t) + v_2(t)$ has Fourier transform $V(\omega) = V_1(\omega) + V_2(\omega)$. This follows directly by substitution into equation 2.29.

(vi) *Time delay:* if $v(t)$ has Fourier transform $V(\omega)$ then the delayed signal

$$v_d(t) = v(t - t_d)$$

has Fourier transform

$$V_d(\omega) = e^{-j\omega t_d} V(\omega) \qquad 2.35$$

The proof follows from the definition of $V_d(\omega)$

$$V_d(\omega) = \int_{-\infty}^{\infty} v(t - t_d) e^{-j\omega t} \, dt$$

$$= e^{-j\omega t_d} \int_{-\infty}^{\infty} v(t - t_d) e^{-j\omega(t - t_d)} \, dt$$

substituting $u = t - t_d$ gives

$$V_d(\omega) = e^{-j\omega t_d} \int_{-\infty}^{\infty} v(u) e^{-j\omega u} \, du$$

$$= e^{-j\omega t_d} V(\omega) \qquad 2.36$$

Time delay thus leaves the modulus of the Fourier transform unchanged but introduces a phase change $\psi = -\omega t_d$, which is a lag proportional to frequency (see also Section 2.6 above).

(vii) *Reciprocal spreading:* if $v(t)$ has Fourier transform $V(\omega)$ then $v(kt)$ has transform $1/k \ V(\omega/k)$.

Writing $V_1(\omega)$ for the transform of $v(kt)$, then, by definition

$$V_1(\omega) = \int_{-\infty}^{\infty} v(kt) e^{-j\omega t} \, dt$$

$$= \int_{-\infty}^{\infty} v(kt) e^{-j(\omega/k) \cdot kt} \, dt$$

$$= \frac{1}{k} \int_{-\infty}^{\infty} v(u) e^{-j\omega u/k} \, du$$

where $u = kt$. Hence

$$V_1(\omega) = \frac{1}{k} V\left(\frac{\omega}{k}\right) \qquad 2.37$$

Note that the signal $v(kt)$ may be thought of as being formed by compressing $v(t)$ in time by a factor k. The Fourier transform becomes stretched out (and reduced in size) by the same factor k—a behaviour termed 'reciprocal spreading'.

2.12 SPECTRA OF SOME SIMPLE PULSES

Some useful Fourier transforms are shown in Fig. 2.14. Transform A for a rectangular pulse of height V and duration τ may be derived in a straightforward manner from equation 2.29

$$V(\omega) = \int_{-\infty}^{\infty} v(t)\,e^{-j\omega t}\,dt$$

$$= V \int_{\tau/2}^{\tau/2} e^{-j\omega t}\,dt$$

$$= V \frac{e^{j\omega\tau/2} - e^{-j\omega\tau/2}}{j\omega}$$

$$= V\tau \,\text{sinc}\,(\omega\tau/2) \qquad\qquad 2.38$$

This result is in agreement with equation 2.26, recalling that $V(\omega)$ was defined as $2\pi g(\omega)$. The pulse in this example has even symmetry and the transform is real and even, as expected from property (ii) above. A consideration of the way in which the shape of $V(\omega)$ in equation 2.38 varies with τ illustrates the reciprocal spreading behaviour of property (vii). Equation 2.33 is easily seen to hold, while equation 2.34 yields the useful definite integral

$$\int_{-\infty}^{\infty} V\tau \,\text{sinc}\,(\omega\tau/2)\,d\omega = 2\pi V$$

or, more generally

$$\int_{-\infty}^{\infty} \text{sinc}\,ax\,dx = \frac{\pi}{a} \qquad\qquad 2.39$$

Transforms B–F are left as problems for the reader. Transform B is easily obtained from transform A using property (iii) of the previous section. The double pulse C may be regarded as the superposition of two rectangular pulses of opposite polarity displaced in time; this transform may therefore be found from that of A using properties (v) and (vii). Note that transform C is purely imaginary, since pulse C has odd symmetry. The remainder may be found by direct application of equation 2.29 (see also Problem 2.7 for pulse D).

37

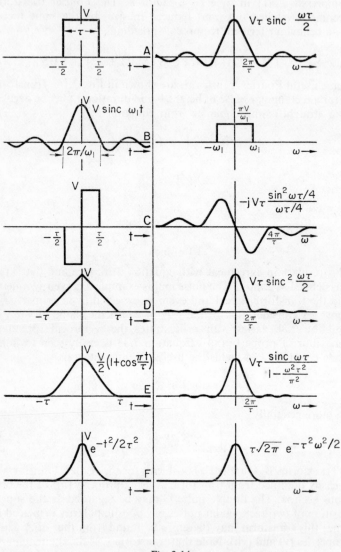

Fig. 2.14.

Pulse F and its transform are remarkable in that they both have the same Gaussian shape and this transform finds an interesting application in probability theory (Section 5.12). Application of equation 2.34 to transform D also yields a definite integral which will be found useful in later work, namely

$$\int_{-\infty}^{\infty} \text{sinc}^2(ax)\,dx = \frac{\pi}{a} \qquad 2.40$$

It should be observed that the pulses in Fig. 2.14 which exist for only a finite time have Fourier transforms which, except at special points, are non-zero at all finite frequencies. A pulse of finite duration, whatever its shape, has an infinite spectrum. Conversely, a pulse whose spectrum is restricted to a finite frequency range (e.g. pulse B) lasts for all time. It is not possible to have a pulse of finite duration whose spectrum extends to only a finite frequency. When signalling with pulses of a given pulse length, however, it is helpful to minimise the high-frequency content as far as possible in order to keep the required channel bandwidth to a minimum. Pulses A, D, and E have been drawn with the same width, τ at half-height. Notice that in each case $V(\omega)$ has its first zero at the same frequency but that the outer lobes of the spectrum of E are much reduced. It has already been observed, in Section 2.4, that rounded waveforms have few significant high-frequency components. Some approximation to the cosine-squared shape of E may thus be usefully employed for signalling. This particular pulse shape also finds an application in testing wide-band amplifiers and is sometimes called a T-pulse.

The choice of τ as the width at half-height of the pulses A, D, and E means that the area under each pulse is the same, namely $V\tau$. If, for these pulses, V is made indefinitely large while simultaneously τ is reduced such that $\tau = k/V$ then in each case we are left with an impulse signal of strength k. In the frequency domain the reciprocal spreading property results in the Fourier transforms of those pulses being stretched out to become the horizontal lines $V(\omega) = k$, as may be verified by allowing $\tau \to 0$ in each of the expressions for $V(\omega)$. This important result, shown as transform G in Fig. 2.15, means that an impulse signal contains frequency components which are uniformly distributed along the whole of the frequency axis (cf. equation 2.9). A similar procedure may be applied to 'pulse' B using equation 2.40 to determine the impulse strength. The converse result, using property (iii) of Section 2.11, is that a d.c. signal V has

4

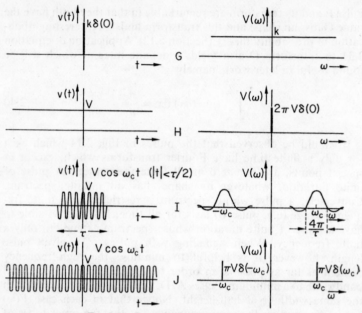

Fig. 2.15. Some further Fourier transforms

as its Fourier transform an impulse function of strength $2\pi V$ at $\omega = 0$. This may be verified by examining the behaviour of the transform B as V sinc ωt is 'stretched out' along the time axis.

The signal

$$v(t) = V\cos\omega_c t \quad -\tau/2 < t < \tau/2$$
$$= 0 \qquad\qquad |t| > \tau/2$$

i.e., a pulse of sinusoidal oscillations lasting for a time τ is sketched as I in Fig. 2.15. The transform $V(\omega)$ is given by

$$V(\omega) = V\int_{-\tau/2}^{\tau/2} \cos\omega_c t \; \mathrm{e}^{-j\omega t}\,\mathrm{d}t$$

$$= \frac{V}{2}\int_{-\tau/2}^{\tau/2} [\mathrm{e}^{j(\omega_c-\omega)t} + \mathrm{e}^{-j(\omega_c+\omega)t}]\,\mathrm{d}t$$

$$= \frac{V\tau}{2}\,[\mathrm{sinc}\,(\omega-\omega_c)\tau/2 + \mathrm{sinc}\,(\omega+\omega_c)\tau/2] \qquad 2.41$$

which contains two identical sinc functions, one centred at $\omega = \omega_c$

40

and the other at $\omega = -\omega_c$ as shown in the figure. The area under each of these sinc functions is easily shown from equation 2.39 to be πV. Allowing $\tau \to \infty$ we therefore obtain the Fourier transform of a *continuous* sine wave as a pair of δ-functions each of strength πV lying at $\omega = \pm\omega_c$. This result, transform **J** in Fig. 2.15, might have been anticipated, for a continuous sine wave $V \cos \omega_c t$ has α-coefficients $V/2$ at $\pm\omega_c$. An α-coefficient (returning to the string-of-beads analogy of Section 2.10) may be regarded as a point 'mass' and can only be represented in terms of a *density* by using a δ-function whose strength equals the mass at the point. The spectral density $g(\omega)$ for $V \cos \omega_c t$ would thus comprise two δ-functions of strength $V/2$ at $\pm\omega_c$ and, recalling that $V(\omega) = 2\pi g(\omega)$, transform **J** follows immediately.

Notice that, although a continuous sinusoid has its whole energy concentrated at one frequency, any sinusoidal signal must in practice be of finite duration and therefore be 'smeared out' in the frequency domain as in Fig. 2.15I. The width of the principal maximum (i.e. between the first zeros of sinc $\omega\tau/2$) is $\Delta\omega = 4\pi/\tau$, so that the shorter τ the greater the spread in frequency (reciprocal spreading again). The pulse I might be used in practice for signalling a Morse dot and we see that a c.w. Morse transmission has a spectrum which will extend to about τ^{-1} Hz either side of the carrier frequency, where τ is the duration of the dot. The faster the message is transmitted (shorter τ) the larger the amount of frequency space (channel bandwidth) which must be allocated to the signal. The rectangular shape of the envelope of the pulse means that, for signalling purposes, an undesirably large amount of energy is 'splashed out' into the side lobes and attention must be paid to the shape of the pulse if this is to be avoided.

2.13 CONVOLUTION, PARSEVAL'S THEOREM, ENERGY SPECTRUM

The Fourier transform of a signal $x(t)$ which is the product of the two signals $v_1(t)$ and $v_2(t)$ may be found in terms of the transforms $V_1(\omega)$ and $V_2(\omega)$ as follows

$$x(t) = v_1(t)\,v_2(t)$$

$$X(\omega) = \int_{-\infty}^{\infty} v_1(t)\,v_2(t)\,\mathrm{e}^{-j\omega t}\,\mathrm{d}t$$

$$= \int_{-\infty}^{\infty} v_1(t)\left[\frac{1}{2\pi}\int_{-\infty}^{\infty} V_2(u)\,\mathrm{e}^{jut}\,\mathrm{d}u\right]\mathrm{e}^{-j\omega t}\,\mathrm{d}t$$

where in replacing $v_2(t)$ by the inverse transform of equation 2.30 we have been careful to introduce the variable u to avoid confusion with ω. Hence

$$X(\omega) = \frac{1}{2\pi} \int_{-\infty}^{\infty} \int_{-\infty}^{\infty} v_1(t)\, V_2(u)\, e^{-j(\omega - u)t}\, du\, dt$$

The order of integration is unimportant here, so that

$$X(\omega) = \frac{1}{2\pi} \int_{-\infty}^{\infty} V_2(u) \left[\int_{-\infty}^{\infty} v_1(t)\, e^{-j(\omega - u)t}\, dt \right] du$$

$$= \frac{1}{2\pi} \int_{-\infty}^{\infty} V_1(\omega - u)\, V_2(u)\, du$$

If we had substituted for $v_1(t)$ instead of $v_2(t)$ in the first step of this derivation, the last line would read

$$X(\omega) = \frac{1}{2\pi} \int_{-\infty}^{\infty} V_1(u)\, V_2(\omega - u)\, du$$

Hence

$$\int_{-\infty}^{\infty} v_1(t)\, v_2(t)\, e^{-j\omega t}\, dt = \frac{1}{2\pi} \int_{-\infty}^{\infty} V_1(u)\, V_2(\omega - u)\, du$$

$$= \frac{1}{2\pi} \int_{-\infty}^{\infty} V_1(\omega - u)\, V(u)\, du \qquad 2.42$$

The integrals on the right-hand side of equation 2.42 are termed *convolution integrals;* thus *the Fourier transform of the product of two signals is given by the convolution of their transforms.*

Conversely, we may ask for the signal whose transform is the product of two Fourier transforms $V_1(\omega)\, V_2(\omega)$. This time the starting point is equation 2.30 into which we substitute for one of the transforms using equation 2.29 with a dummy variable τ. The argument runs precisely parallel to that given above and yields

$$\frac{1}{2\pi} \int_{-\infty}^{\infty} V_1(\omega)\, V_2(\omega)\, e^{j\omega t}\, d\omega = \int_{-\infty}^{\infty} v_1(\tau)\, v_2(t - \tau)\, d\tau$$

$$= \int_{-\infty}^{\infty} v_1(t - \tau)\, v_2(\tau)\, d\tau \qquad 2.43$$

where the convolutions are in the time domain.

Equation 2.43 is important in the theory of linear filtering and will be discussed again, together with a graphical picture of the convolution integral in Section 3.11. Meanwhile, equation 2.42 is of more immediate interest, for with $\omega = 0$ and using property (i) of Section 2.11 we have

$$\int_{-\infty}^{\infty} v_1(t)\,v_2(t)\,\mathrm{d}t = \frac{1}{2\pi}\int_{-\infty}^{\infty} V_1(u)\,V_2^*(u)\,\mathrm{d}u$$

$$= \frac{1}{2\pi}\int_{-\infty}^{\infty} V_1(\omega)\,V_2^*(\omega)\,\mathrm{d}\omega \qquad 2.44$$

which is a general form of Parseval's theorem. For the special case $v_1(t) \equiv v_2(t) \equiv v(t)$

$$\int_{-\infty}^{\infty} v^2(t)\,\mathrm{d}t = \frac{1}{2\pi}\int_{-\infty}^{\infty} |V(\omega)|^2\,\mathrm{d}\omega$$

$$= \frac{1}{\pi}\int_{0}^{\infty} |V(\omega)|^2\,\mathrm{d}\omega \qquad 2.45$$

The left-hand side of equation 2.45 represents the total *energy* delivered (to a $1\,\Omega$ resistor) by the signal $v(t)$. This energy may thus be computed from the Fourier transform. Equation 2.45 is formally similar to the expressions obtained in equations 2.15 and 2.25 for the mean *power* of a periodic signal. As an example, the Fourier transform for pulse A of Fig. 2.14 may be inserted in equation 2.45. The integral is evaluated using 2.40, giving $V^2\tau$ for the energy of the pulse, in agreement with the time-domain calculation.

It is possible to interpret equation 2.45 in the following way: given a signal $v(t)$ having Fourier transform $V(\omega)$ the energy carried by components in the frequency range ω_1 to ω_2 is

$$E(\omega_1, \omega_2) = \frac{1}{\pi}\int_{\omega_1}^{\omega_2} |V(\omega)|^2\,\mathrm{d}\omega \qquad 2.46$$

Thus, the fraction η of the total pulse energy which lies in the frequency range from d.c. to ω is given by

$$\eta = \frac{\displaystyle\int_{0}^{\omega} |V(\omega)|^2\,\mathrm{d}\omega}{\displaystyle\int_{0}^{\infty} |V(\omega)|^2\,\mathrm{d}\omega} \qquad 2.47$$

This function is plotted in Fig. 2.16 for the rectangular and cosine-

43

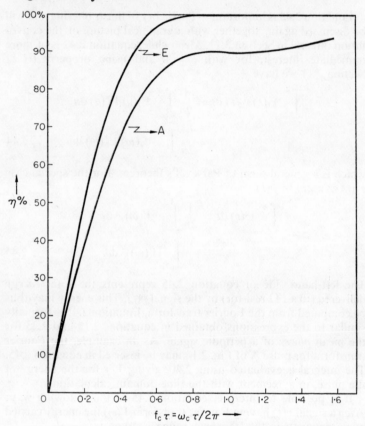

Fig. 2.16. Fraction of the total energy in pulses A and E of Fig. 2.14 contained within the frequency range 0–f_c Hz

squared pulses of Fig. 2.14 A and E. Equation 2.46 shows that $|V(\omega)|^2$ may be regarded as a density distribution of the pulse energy along the frequency axis and may be termed the *energy spectrum* of the pulse.

2.14 SIGNALLING RATE AND BANDWIDTH

Consider a binary signalling system in which a pulse of duration τ is used to signify *mark* (or 1) and the absence of a signal for the same period is used to signify *space* (or 0). It is clear from inspection

44

of Fig. 2.16 that, as the bandwidth of a communication link must be finite, less than 100% of the transmitted signal energy will be received, even if the attenuation of the link within the available bandwidth is negligible. Some distortion of the pulse shape will also occur, and if the bandwidth is too small the received waveform will be a meaningless jumble in which the pulse sequences cannot be discerned. This kind of distortion will be examined in the following chapter, but for the moment we may suppose that, provided the bandwidth extends to at least f_{min} Hz, then a sufficient fraction, say $\eta = 90\%$, of the signal energy is received and distortion is not serious. Now Fig. 2.16 shows that for a given η the value of $f_{min}\tau$ is fixed at a value which depends on the shape of the pulses used. If the signalling rate is defined as the number of 'ones' or 'zeros' transmitted per second then

$$\text{signalling rate} = \frac{1}{\tau} \propto f_{min}$$

So the maximum signalling rate possible is proportional to the bandwidth of the link. If pulses of carrier waves are used, the discussion in Section 2.12 relating to transform I of Fig. 2.15 shows that the same result applies, where in this case the bandwidth is measured around the carrier frequency.

The conclusion above is not restricted to binary signalling systems, as the following example will show. Consider any message waveform (e.g. a speech waveform) lasting for time T. An operator may attempt to transmit this message in time $T/2$ (thereby achieving double the rate of information transfer) by making a tape recording of the waveform and playing this over the communication link with the tape running at twice the speed used for recording. This process may be reversed at the receiving end—time is wasted, of course, in the record–playback operations but the actual time the link is in use is $T/2$. This procedure, however, compresses the waveform in the time domain by a factor of two and therefore the spectrum is stretched out by the same factor. A signal transmitted at twice the speed occupies twice the bandwidth.

2.15 BAND-LIMITED SIGNALS.
SAMPLING THEOREM

A band-limited signal is one for which $V(\omega) \equiv 0$ for $|\omega| > W$, i.e. the spectrum is non-zero only up to some limiting angular frequency $W = 2\pi F$. As was pointed out in Section 2.12 above, such

45

a signal must exist for all time, though it may in practice be regarded as negligible outside a certain time interval. The special property of such signals is that they may be described in the time domain by quoting $v(t)$ not as a continuous sequence of values but as a discrete sequence of the values taken by $v(t)$ at intervals t_s. Given the *sample values* v_k taken at the *sampling instants* $t_k = t_0 + kt_s$, where t_0 is arbitrary and k is any positive or negative integer, we have, effectively, a sequence of points on the graph $v(t)$ against t. The sampling theorem states that, provided $t_s \leqslant 1/2F$, these samples specify the band-limited signal $v(t)$ completely. The graph joining the points may be drawn unambiguously, without any doubts about how to interpolate between them.

An elementary justification of the theorem argues as follows. Suppose $v(t)$ is a periodic waveform, with fundamental interval $-T/2 < t < T/2$ so that

$$v(t) = A_0 + \sum_{n=1}^{N} A_n \cos n\omega_1 t + \sum_{n=1}^{N} B_n \sin n\omega_1 t$$

where the range of n is no longer infinite but extends only up to N so that $v(t)$ is band-limited to an upper frequency $W = N\omega_1$. If we were told the values of the $2N+1$ coefficients, A_0, A_n, and B_n, then $v(t)$ would be known for any t. Alternatively, we might be told $2N+1$ sample values of $v(t)$ at $2N+1$ different sampling instants t_k within the fundamental interval. Substitution in the equation above would then provide $2N+1$ simultaneous equations from which the $2N+1$ Fourier coefficients could be calculated. Assuming a uniform sampling rate, the sampling interval must be

$$t_s = \frac{T}{2N+1}$$

$$= \frac{T}{(WT/\pi)+1}$$

The sampling procedure is most useful for long, complicated signals for which $WT \gg 1$, letting $WT \to \infty$ in the above equation

$$t_s = \frac{\pi}{W} = \frac{1}{2F} \qquad 2.48$$

which represents the maximum permitted sampling interval.

In general terms the theorem is in agreement with intuition, for a signal whose most rapidly varying component lies at F Hz cannot pass through maxima and minima more rapidly than $2F$ times per second. There is thus no question of the graph of $v(t)$ wandering about indeterminately between the sample points, which must therefore be joined up by a smooth curve.

A rather more general proof of the theorem, which also provides a much simpler interpolation procedure, begins with the Fourier transform $V(\omega)$. Within a fundamental interval, $-W < \omega < W$, this function may itself be written as a complex Fourier series

$$V(\omega) = \sum_{n=-\infty}^{\infty} \alpha_n e^{j(n\pi\omega)/W} \qquad 2.49$$

where

$$\alpha_n = \frac{1}{2W} \int_{-W}^{W} V(\omega)\, e^{-j(n\pi\omega)/W}\, d\omega \qquad 2.50$$

Substituting $V(\omega)$ from equation 2.49 into 2.30 we have

$$v(t) = \sum_{n=-\infty}^{\infty} \left\{ \frac{1}{2\pi} \int_{-W}^{W} \alpha_n e^{j(n\pi\omega)/W}\, e^{j\omega t}\, d\omega \right\}$$

$$= \sum_{n} \alpha_n \left\{ \frac{1}{2\pi} \int_{-W}^{W} e^{j\omega(t + n\pi/W)}\, d\omega \right\}$$

where the range of integration is restricted to $\pm W$; within this range equation 2.49 is valid, outside this range $V(\omega) \equiv 0$, so the integral is equivalent to equation 2.30 with infinite limits. The integral within the bracket yields

$$\frac{1}{2\pi} \int_{-W}^{W} e^{j\omega(t + n\pi/W)}\, d\omega = \frac{W}{\pi}\ \text{sinc}\ W(t + n\pi/W)$$

Hence

$$v(t) = \sum_{n=-\infty}^{\infty} \frac{\alpha_n W}{\pi}\ \text{sinc}\ W(t + n\pi/W)$$

Finally, comparing equation 2.50 with 2.30

$$\alpha_n = \frac{\pi}{W}\, v\!\left(-\frac{n\pi}{W}\right) \qquad 2.51$$

47

hence

$$v(t) = \sum_{n=-\infty}^{\infty} v\left(-\frac{n\pi}{W}\right) \text{ sinc } W(t+n\pi/W)$$

or, writing $k = -n$ and $\pi/W = t_s$

$$v(t) = \sum_{k=-\infty}^{\infty} v_k \text{ sinc } W(t-kt_s) \qquad\qquad 2.52$$

This equation expresses the band-limited signal $v(t)$ in terms of its sample values v_k taken at the instants $t = kt_s$, where $t_s = \pi/W = 1/2F$. Now, any signal which is band-limited to W is also band-limited to W_1 where $W_1 \geqslant W$. The whole of the foregoing analysis may therefore be carried through with W_1 replacing W, in which case the sampling interval t_s would have been smaller. The argument becomes invalid, however, for $W_1 < W$, hence

$$t_s \leqslant 1/2F \qquad\qquad 2.53$$

The sampling theorem may therefore be stated as follows: *Any signal which is band-limited to an upper frequency limit F Hz is completely specified by stating its values at a rate 2F per second or faster.* The minimum sampling rate is called the Nyquist rate. (The idea that a signal limited to 0–F Hz transmits 2F sample values per second is attributed to Nyquist.[3] The form expressed by equation 2.52 is given by C. E. Shannon, *Proc. Inst. Radio Engrs* **37**, 10–21 (1949).)

Equation 2.52 gives a method of reconstituting the signal from its sample values. On each sampling instant construct a function sinc Wt with a central amplitude equal to the sample value. The signal $v(t)$ is the linear superposition of these components. This procedure is sketched in Fig. 2.17; note that the function sinc Wt centred on a given sampling instant is zero at all other sampling instants. Between the sample points, however, the signal $v(t)$ depends on the value of every sample from $t = -\infty$ to $t = \infty$. Because sinc Wt decays slowly with time, only those samples lying close to some chosen instant t' will usually be important in contributing to $v(t')$ but, given only the samples from a finite section of $v(t)$, a restricted sum of the form of equation 2.52 cannot be expected to give a precise replica of the chosen section, especially near the ends.

The theorem may also be looked at the other way round. If a received signal waveform, band-limited to F Hz, is sampled $2F$ times per second, each of these samples is independent of the others. Each

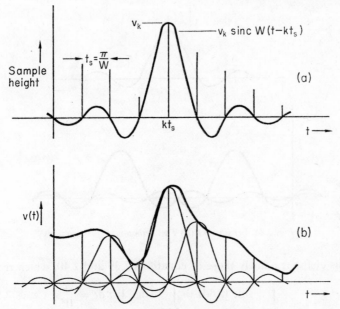

Fig. 2.17. Illustrating equation 2.52

sample gives a fresh piece of information about the signal and, since these pieces of information arrive at a rate proportional to the upper frequency limit, we have a further illustration of the theme of Section 2.14, that the maximum rate of information transfer is proportional to the bandwidth used.

2.16 PROPERTIES OF THE SIGNAL sinc Wt AND SOME DERIVED RESULTS

Some of the more important properties of the function sinc Wt are gathered together in this section, for convenience of reference. The opportunity will also be taken to derive expressions for the mean and mean square value of a band-limited signal in terms of its sample values.

The function sinc Wt takes the value unity at $t = 0$. The interval between the zeros on either side of this principal maximum is $2\pi/W$, but elsewhere the zeros are uniformly spaced, at intervals π/W (Fig. 2.18(a)). Two definite integrals have been evaluated earlier in

49

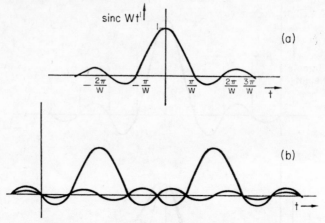

Fig. 2.18. (a) *sinc Wt* (b) *Two orthogonal sinc functions*

this chapter, namely those of equations 2.39 and 2.40, which read

$$\int_{-\infty}^{\infty} \text{sinc } Wt \text{ d}t = \int_{-\infty}^{\infty} \text{sinc}^2 Wt \text{ d}t = \frac{\pi}{W} \qquad 2.54$$

The indefinite integral cannot be evaluated analytically, although tables exist for the function $Si(x)$ which is defined as

$$Si(x) = \int_{0}^{x} \text{sinc } u \text{ d}u \qquad 2.55$$

This function will appear in the following chapter as describing the response of a low-pass filter to an input step-function (Section 3.5).

The two signals

$$v_1(t) = \text{sinc } W(t - n\pi/W)$$
$$v_2(t) = \text{sinc } W(t - m\pi/W)$$

have, using transform B of Fig. 2.14 and the time delay theorem of Section 2.11, the Fourier transforms

$$V_1(\omega) = \frac{\pi}{W} e^{-jn\pi\omega/W} \qquad |\omega| < W$$

$$V_2(\omega) = \frac{\pi}{W} e^{-jm\pi\omega/W} \qquad |\omega| < W$$

$$V_1(\omega) = V_2(\omega) \equiv 0 \qquad |\omega| > W$$

Equation 2.44 now gives

$$\int_{-\infty}^{\infty} v_1(t)\,v_2(t)\,\mathrm{d}t = \frac{1}{2\pi}\int_{-\infty}^{\infty} V_1(\omega)\,V_2^*(\omega)\,\mathrm{d}\omega$$

$$= \frac{\pi}{2W^2}\int_{-W}^{W} \mathrm{e}^{j(m-n)\pi\omega/W}\,\mathrm{d}\omega$$

$$= 0, \quad m \neq n.$$

Any two sinc pulses which are separated in time by an integer multiple of their zero crossing interval π/W, as in Fig. 2.18(b), are therefore orthogonal. It is now seen that equation 2.52 expresses a band-limited signal as a superposition of mutually orthogonal component signals in a manner similar to the Fourier expansion, equation 2.3.

The average value of a band-limited signal $v(t)$ over a time T which contains many samples may be transformed using equation 2.52 to yield

$$\overline{v(t)} = \frac{1}{T}\int_{-T/2}^{T/2} v(t)\,\mathrm{d}t = \frac{1}{T}\sum_{k=-\infty}^{\infty} v_k \int_{-T/2}^{T/2} \mathrm{sinc}\,W(t-k\pi/W)\,\mathrm{d}t$$

If $WT \gg 1$ then the number of samples, $N = WT/\pi$ contained in the interval T is large. Little error will be involved if the contributions from the sample functions outside this range are neglected. The range of k may therefore be restricted $k = -WT/2\pi$ to $k = WT/2\pi$. A second approximation, again valid for $WT \gg 1$, is to replace the limits of integration by $\pm\infty$, thus

$$\overline{v(t)} = \frac{1}{T}\sum_{k=-WT/2\pi}^{WT/2\pi} v_k \int_{-\infty}^{\infty} \mathrm{sinc}\,W(t-k\pi/W)$$

$$= \frac{\pi}{WT}\sum_{k} v_k = \frac{N\overline{v_k}\pi}{WT}$$

where $N = WT/\pi$ is the number of samples in T. Hence

$$\overline{v(t)} = \overline{v_k} \quad (WT \gg 1) \qquad 2.56$$

i.e. the mean value of $v(t)$ is the mean of the sample values.

Similarly, when $v(t)$ from equation 2.52 is substituted into the expression for the mean square value

$$\overline{v^2(t)} = \frac{1}{T}\int_{-T/2}^{T/2} v^2(t)\,\mathrm{d}t$$

the cross-product terms vanish on integration, since the component sample functions are orthogonal for $WT \rightarrow \infty$, leaving

$$\overline{v^2(t)} = \frac{1}{T} \sum_k v_k^2 \int_{-\infty}^{\infty} \text{sinc}^2\, W(t - k\pi/W)\, \mathrm{d}t$$

$$= \frac{\pi}{WT} \sum_k v_k^2$$

$$= \overline{v_k^2} \qquad\qquad 2.57$$

Thus, the mean power of a band-limited signal may also be computed from the sample values. These results will be useful when dealing with noise waveforms in Chapter 7.

2.17 SPECTRUM OF A TRAIN OF PULSES

If $v(t)$ is a pulse of arbitrary shape Fig. 2.19(a) then the signal

$$x(t) = \sum_k v(t - kT)$$

represents a train of pulses repeating at intervals T (Fig. 2.19(b)).

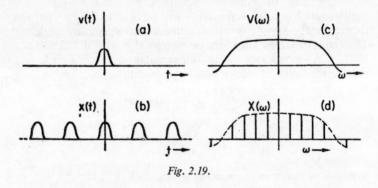

Fig. 2.19.

The pulses may or may not overlap. Applying properties (v) and (vi) of Section 2.11 we have

$$X(\omega) = V(\omega)\,[1 + e^{j\omega T} + e^{-j\omega T} + e^{j2\omega T} + e^{-j2\omega T} + \ldots]$$

$$= 2V(\omega)\,[\tfrac{1}{2} + \cos \omega T + \cos 2\omega T + \ldots]$$

Comparing this with equation 2.9, the expression within the square brackets is seen to describe a train of δ-functions along the ω-axis

$$X(\omega) = \frac{2\pi V(\omega)}{T} \sum_n \delta(\omega - 2\pi n/T) \qquad 2.58$$

Thus, the Fourier transform of a pulse train has the form of a row of δ-functions lying at the harmonics of the pulse repetition frequency, the strengths of the δ-function being governed by the transform of the individual pulses. The spectra of $v(t)$ and $x(t)$ are represented in Figs 2.19(c) and (d).

Reverting briefly to the spectral density $g(\omega)$ of Section 2.10

$$g(\omega) = \frac{X(\omega)}{2\pi} = \frac{V(\omega)}{T} \sum_n \delta(\omega - 2\pi n/T)$$

but each δ-function in $g(\omega)$ represents a complex Fourier coefficient α, (see discussion following equation 2.41). Thus

$$\alpha_n = \frac{V(\omega)}{T}, \quad \text{where} \quad \omega = \frac{2\pi n}{T} \qquad 2.59$$

As an example, consider the train of pulses sinc Wt with repetition interval T as shown in Fig. 2.20(a). Inserting transform B of

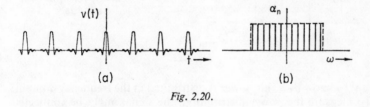

(a)   (b)

Fig. 2.20.

Fig. 2.14 into equation 2.59 yields

$$\alpha_n = \frac{\pi}{WT}, \quad |n| < WT/2\pi$$

$$\alpha_n = 0, \quad |n| > WT/2\pi$$

This spectrum comprises a uniform array of α-coefficients in the range $-W < \omega < W$ as shown in Fig. 2.20(b). A similar technique may be used to find the spectrum of a train of rectangular pulses and the reader is invited to verify equation 2.24 using this method.

2.18 RANDOM WAVEFORMS

Imagine a binary-coded message in which a signal level V lasting for time τ is used to transmit a 1 and $-V$ is used to transmit a 0. Each message transmitted using this arrangement will have its corresponding signal waveform, some section of which might look like that shown in Fig. 2.21. Clearly, the power of the signal is V^2, but in designing circuits to handle the signal it would be impor-

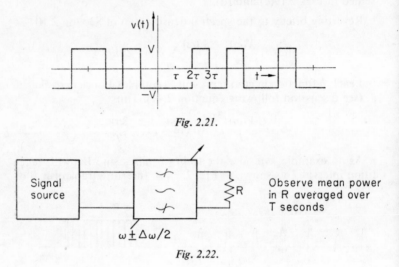

Fig. 2.21.

Fig. 2.22.

tant to know how this power is distributed in the frequency domain. To measure the power distribution, the signal might be connected to a load via a bandpass filter of variable centre frequency ω and bandwidth $\Delta\omega$ as in Fig. 2.22. In the time interval from $-T/2$ to $T/2$ the total energy within $\Delta\omega$ delivered by the signal, divided by T, would represent the mean power associated with the frequency range $\omega \pm \Delta\omega/2$. This power would be a function of ω but would also be different for every message. If T were made sufficiently long, however, then it is reasonable to suppose that the mean power received in $\Delta\omega$ would tend to some well-defined average value which would be the same for every (long) message. We are obliged to deal with average quantities, for we are dealing with what is essentially a *random waveform*. At each of the moments τ, 2τ, 3τ, etc., the signal may or may not change sign. As regards average power distribution, the particular message is unimportant and the $+V$ or $-V$ choices may be regarded as random. The waveform generated by a television camera covering a 'live' scene is also random in this sense, as opposed to the deterministic waveform produced when the camera is focussed on a test-chart. The latter waveform is exactly repetitive, having its power concentrated at the discrete Fourier harmonic frequencies as indicated by equation 2.15. The frequency distribution of the power of a periodic waveform is a line spectrum, while that of a random waveform is continuous. Other examples of random waveforms are those generated by random noise processes and will be discussed in Chapters 6 and 7 (see Fig. 6.1). The purpose of this section is to develop the frequency domain concept of a power density spectrum for random waveforms.

If the filter bandwidth $\Delta\omega$ in Fig. 2.22 is made sufficiently small, and the averaging time T is made sufficiently long, then this experiment may be regarded as a measurement of the *power density spectrum* $P(\omega)$. Writing ΔP as the mean power received between $\omega - \Delta\omega/2$ and $\omega + \Delta\omega/2$, we may write

$$\Delta P = P(\omega)\,\Delta\omega$$

where $P(\omega)$ is a power spectral density measured in $V^2\,\mathrm{rad}^{-1}\,\mathrm{s}$. In this definition ω is a positive number and $P(\omega)$ is said to be 'one-sided'. Analytically, it will be more convenient here to deal with a two-sided power spectrum $p(\omega)$ in which ω may be positive or negative such that

$$p(-\omega) = p(\omega)$$

and also to change the units so that $p(\omega) = p(2\pi f)$ represents the power density in $V^2\,\mathrm{Hz}^{-1}$ at $\omega = 2\pi f$. In a practical observation there will be equal contributions from both positive and negative parts of the frequency axis, thus

$$\Delta P = 2p(2\pi f)\,\Delta f = \frac{p(\omega)}{\pi}\,\Delta\omega \qquad\qquad 2.60$$

represents the observed power in the range $\omega \pm \Delta\omega/2$.

An insight into the nature of the power density spectrum may be obtained by writing down the Fourier series expansion for a section of a random signal from $t = -T/2$ to $t = +T/2$,

$$v(t) = \sum_{n=-\infty}^{\infty} \alpha_n e^{jn\omega_1 t} \qquad |t| < T/2$$

The power associated with the coefficient α_n at angular frequency $n\omega_1$ is (equation 2.25)

$$\frac{\Delta P}{2} = |\alpha_n|^2 = \left| \frac{1}{T} \int_{-T/2}^{T/2} v(t)\, e^{-jn\omega_1 t}\, \mathrm{d}t \right|^2$$

If we now allow T to become very large, and simultaneously allow n to increase, keeping $\omega = n\omega_1$ constant, then the power associated with the interval $\omega \pm \omega_1/2$ becomes

$$\frac{\Delta P}{2} = |\alpha_\omega|^2 = \left| \frac{1}{T} \int_{-T/2}^{T/2} v(t)\, e^{-j\omega t}\, \mathrm{d}t \right|^2$$

One might proceed and write for the *power density*

$$p(\omega) = \frac{\pi\, \Delta P}{\omega_1} = T|\alpha_\omega|^2 = \frac{1}{T} \left| \int_{-T/2}^{T/2} v(t)\, e^{-j\omega t}\, \mathrm{d}t \right|^2 \qquad 2.61$$

in the hope that this quantity will tend to some well-defined limit as $T \to \infty$. If, however, T is made ten times longer, then the 'new' α_ω need bear no simple relationship to the old one, for it relates to a substantially different waveform. This will always be true, no matter what the value of T, and it cannot be expected that a limit of this suggested form does in fact exist. The quantity $T|\alpha_\omega|^2$ is really a random variable (see Section 5.5) and it is the mean value of this which ought to be considered. One way to define the power density for the random telegraph signal considered above Fig. 2.21 would be to imagine a large collection (or *ensemble*) of such waveforms and to evaluate the quantity $T|\alpha_\omega|^2$ for a length T of each. The mean value of the numbers so obtained does approach a definite limit as $T \to \infty$ and the power density may be written

$$p(\omega) = \lim_{T \to \infty} T\, \overline{|\alpha_\omega|^2} \qquad 2.62$$

where the bar signifies that an average has been taken over the ensemble. In a practical measurement of power spectral density using a filter of bandwidth $\Delta\omega$ in the way suggested by Fig. 2.22, the averaging time T must be long compared with the reciprocal of the bandwidth, for if this is the case

$$T \gg \frac{1}{\Delta\omega}$$

i.e. $$\Delta\omega \gg \omega_1$$

so that, effectively, the contributions of many α-coefficients are being added together. The result will be an average, not of α-coefficients from different waveforms, but of a number of neighbouring coefficients in a single waveform, but the effect is the same. It means that if a random signal is applied to a wave analyser, a steady deflection of the needle is observed, provided the meter response time (T) is appreciably longer than the reciprocal of the bandwidth of the instrument.

2.19 POWER SPECTRUM AND AUTOCORRELATION

The above discussion, while relating the power spectrum to the Fourier representation of the signal, does not offer a very practical method for its calculation. The power spectrum is, however, closely related to a function which may often be calculated from the statistical properties of the signal. This function is the *autocorrelation function*, defined as

$$R(s) = \lim_{T \to \infty} \frac{1}{T} \int_{-T/2}^{T/2} v(t)v(t+s) \, dt \qquad 2.63$$

i.e., it is the average of the product of sample values taken s seconds apart. It is clear that $R(s)$ depends only on the size of the interval s and not on its sign, thus

$$R(-s) = R(s) \qquad 2.64$$

Setting $s = 0$ in equation 2.63, it is seen that

$$R(0) = \lim_{T \to \infty} \frac{1}{T} \int_{-T/2}^{T/2} v^2(t) \, dt \qquad 2.65$$

is the mean signal power.

Suppose that a 'truncated' signal $v_T(t)$ is now defined as

$$v_T(t) = v(t) \qquad |t| < T/2$$
$$v_T(t) = 0 \qquad |t| > T/2$$

The integral in equation 2.63 may now be written ($s \ll T$),

$$\int_{-T/2}^{T/2} v(t)v(t+s) \, dt = \int_{-\infty}^{\infty} v_T(t)v_T(t+s) \, dt$$

and writing

$$v_1(t) = v_T(t), \qquad v_2(t) = v_T(t+s)$$

so that $\qquad V_1(\omega) = V_T(\omega), \qquad V_2(\omega) = e^{j\omega s} V_T(\omega)$

equation 2.44 reads

$$\int_{-T/2}^{T/2} v(t)\, v(t+s)\, \mathrm{d}t = \frac{1}{2\pi} \int_{-\infty}^{\infty} V_T(\omega)\, V_T^*(\omega)\, e^{-j\omega s}\, \mathrm{d}\omega$$

$$= \frac{1}{2\pi} \int_{-\infty}^{\infty} |V_T(\omega)|^2\, e^{-j\omega s}\, \mathrm{d}\omega$$

$$= \frac{1}{2\pi} \int_{-\infty}^{\infty} |V_T(\omega)|^2\, e^{j\omega s}\, \mathrm{d}\omega \qquad 2.66$$

where in the last step the complex conjugate has been taken on both sides. But from the definition of $v_T(t)$

$$V_T(\omega) = \int_{-T/2}^{T/2} v(t)\, e^{-j\omega t}\, \mathrm{d}t$$

so that, in the sense of equation 2.61

$$|V_T(\omega)|^2 = T p(\omega)$$

Inserting this into equations 2.66 yields

$$\frac{1}{T} \int_{-T/2}^{T/2} v(t)\, v(t+s)\, \mathrm{d}t = \frac{1}{2\pi} \int_{-\infty}^{\infty} p(\omega)\, e^{j\omega s}\, \mathrm{d}\omega$$

and, passing lightly over the problems involved as $T \to \infty$, we have

$$R(s) = \frac{1}{2\pi} \int_{-\infty}^{\infty} p(\omega)\, e^{j\omega s}\, \mathrm{d}\omega \qquad 2.67$$

and hence

$$p(\omega) = \int_{-\infty}^{\infty} R(s) e^{-j\omega s}\, \mathrm{d}s \qquad 2.68$$

The autocorrelation function $R(s)$ and the power density spectrum $p(\omega)$ are seen to be related as Fourier transforms, a result known as the Wiener–Kinchine relation.

By way of illustrating the significance of equations 2.67 and 2.68, the so-far-unsolved problem of finding the power spectrum of the random signal of Fig. 2.21 will be approached by way of the auto-correlation function $R(s)$. To evaluate $R(s)$ it is necessary to find the mean value of the product $v_1 v_2$ where v_1 and v_2 are samples taken s seconds apart. This product may be one of the four possibilities: VV, $V(-V)$, $(-V)V$, or $(-V)(-V)$. Now, when $s > \tau$, the two samples must come from different signalling intervals so that, assuming $+V$ and $-V$ are equally probable, $v_1 v_2$ is equally likely to be V^2 or $-V^2$. Thus

$$R(s) = 0 \qquad |s| > \tau$$

For $s < \tau$ the chance that v_1 lies less than s seconds from the end of a signalling interval is s/τ. Hence a fraction s/τ pairs have v_1 and v_2 in adjacent signalling intervals and for one half of these v_1 and v_2 are of opposite sign. Thus a fraction $s/2\tau$ have $v_1 v_2 = -V^2$ while $(1-s/2\tau)$ have $v_1 v_2 = +V^2$. The mean value is therefore given by

$$R(s) = \overline{v_1 v_2} = V^2(1 - |s|/\tau) \qquad |s| < \tau$$

This result is sketched in Fig. 2.23(a). The Fourier transform of $R(s)$ is transform D of Fig. 2.14. Hence

$$p(\omega) = V^2 \int_{-\tau}^{\tau} (1 - |s|/\tau)\, e^{-j\omega s}\, \mathrm{d}s$$
$$= V^2 \tau\, \mathrm{sinc}^2 (\omega\tau/2) \qquad\qquad 2.69$$

which is sketched in Fig. 2.23(b).

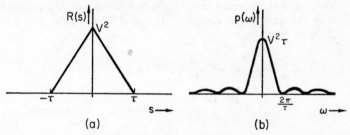

(a)   (b)

Fig. 2.23. Autocorrelation function $R(s)$ and power density spectrum $p(\omega)$ for the random waveform of Fig. 2.21.

The total power is given by

$$P = \frac{1}{2\pi} \int_{-\infty}^{\infty} p(\omega)\, \mathrm{d}\omega = \frac{V^2 \tau}{2\pi} \int_{-\infty}^{\infty} \mathrm{sinc}^2 (\omega\tau/2)\, \mathrm{d}\omega$$
$$= V^2$$

which is also the value of $R(0)$, as expected.

Signals and Information

It happens that, for this particular example, the power spectrum is identical in form to the *energy* density spectrum $|V(\omega)|^2$ of the single rectangular pulses of which the waveform is composed. This, however, is not generally the case. For a fuller discussion on autocorrelation and power spectra the reader is referred to the more advanced texts on communication theory and to reference 4 below.

References

1. GIBBS, J. WILLARD, *Nature, Lond.* **59**, 606 (1899)
2. KRAKAUER, S. M., 'Harmonic generation, rectification and lifetime evaluation with step recovery diodes', *Proc. Inst. Radio Engrs* **50**, 1065 (1962)
3. NYQUIST, H., 'Certain topics in telegraph transmission theory', *Trans. Am. Inst. Elect. Engrs* **47**, No. 2, 617 (1928)
4. BENNETT, W. R., 'Methods of solving noise problems', *Proc. Inst. Radio Engrs* **44**, 609 (1956)

Further Reading

GOLDMAN, S. *Frequency Analysis, Modulation and Noise* (McGraw-Hill, New York, 1948)
HSU, H. P., *Fourier Analysis* (Associated Educational Services Corp., New York, 1967)
LATHI, B. P., *Communication Systems* (Wiley, New York, 1968)

PROBLEMS

1. Show that the Fourier series for a half-wave rectified sinusoid, with the time origin chosen at the peak of a conducting half-cycle, is given by

$$v_1(t) = \frac{1}{\pi} + \frac{1}{2} \cos \omega_1 t + \frac{2}{3\pi} \cos 2\omega_1 t - \frac{2}{15\pi} \cos 4\omega_1 t$$
$$+ \ldots + \frac{2(-1)^{(n/2)+1}}{\pi(n^2-1)} \cos n\omega_1 t + \ldots \quad (n \text{ even})$$

2. Show that if the time origin in (1) is chosen at the start of a conducting half-cycle, the expansion becomes

$$v_2(t) = \frac{1}{\pi} + \frac{1}{2} \sin \omega_1 t - \frac{2}{3\pi} \cos 2\omega_1 t - \frac{2}{15\pi} \cos 4\omega_1 t$$
$$- \ldots - \frac{2}{\pi(n^2-1)} \cos n\omega_1 t - \ldots \quad (n \text{ even})$$

3. Starting from $v_1(t)$ of question (1), sketch and write down the series for the waveform

$$v_3(t) = v_1(t) - \cos \omega_1 t$$

Hence, show that the Fourier series for a full-wave rectified sinusoid (peak at $t = 0$) is given by

$$v_4(t) = \frac{4}{\pi} \left[\frac{1}{2} + \frac{1}{3} \cos 2\omega_1 t - \frac{1}{15} \cos 4\omega_1 t + \frac{1}{37} \cos 6\omega_1 t + \ldots \right]$$

4. Derive the Fourier expansion for the triangular wave of Fig. 2.9(b) using equations 2.4 directly. Evaluate the mean power of this signal (i) in the time domain (ii) in the frequency domain (equation 2.15).

5. Show that approximately 90% of the power in a square wave is carried by the first and third harmonics.

6. Find the Fourier transform of the pulse given by

$$v(t) = \tfrac{1}{2} e^{-at} \quad (t > 0)$$
$$= -\tfrac{1}{2} e^{at} \quad (t < 0) .$$

and hence, by taking the limit as $a \to 0$, show that a unit step has the transform $V(\omega) = 1/j\omega$.

7. Show from equation 2.30 that if $v(t)$ has transform $V(\omega)$, then the signal $v'(t)$ has transform $j\omega \, V(\omega)$. Verify that transforms C and D of Fig. 2.14 are related in this way, and also the transforms for a unit step and unit impulse.

8. Find the transform of the signal $v(t) = \delta(t - t_d) + \delta(t - t_d)$. (Compare your result with the inverse of transform J, Fig. 2.15, using property (iii) of Section 2.11.)

9. Using equations 2.23 and 2.24, show that

$$\sum_{n=-\infty}^{\infty} \text{sinc } n\theta = \frac{\pi}{\theta} \quad (\theta \leq \pi)$$

10. Show that the energy in the pulse $v(t) = V \text{ sinc } Wt$ is $\dfrac{V^2\pi}{W}$ V^2 s

11. By expanding the function $v(t) = \text{sinc } W(t-t_1)$ in the form of equation 2.52, and noting that $v(t_1) = 1$, show that

$$\sum_{n=-\infty}^{\infty} \text{sinc}^2 (Wt_1 - n\pi) = 1$$

12. Show from equation 2.43 that the convolution of two Gaussian pulses

$$v_1(t) = V_1 e^{-at^2} \quad \text{and} \quad v_2(t) = V_2 e^{-bt^2}$$

is Gaussian.

13. Show that a band-limited signal must last for all time. (Hint: let $V_1(\omega)$ be the transform of a band-limited signal and $V_2(\omega)$ be unity within the band and zero elsewhere. The inverse transform of $V_1(\omega) V_2(\omega)$ is $v_1(t)$ but $v_1(t)$ may now also be written as a convolution integral by equation 2.43.)

14. Use the vector representation method of Appendix 3 to show that

$$\left| \int_{-\infty}^{\infty} v_1(t) \, v_2(t) \, \mathrm{d}t \right|^2 \leqslant \left| \int_{-\infty}^{\infty} v_1^2(t) \, \mathrm{d}t \right| \times \left| \int_{-\infty}^{\infty} v_2^2(t) \, \mathrm{d}t \right|$$

(Schwarz's inequality).

Chapter 3

Signal Processing

3.1 INTRODUCTION

In passing through an amplifier or over a long cable, an electrical signal will become distorted if some of its frequency components are amplified or attenuated more than others or if their phase relationships are altered. This distortion will always arise when the signal contains frequency components which lie outside the range of frequencies the system was designed to handle. Usually the bandwidth of a communication link is chosen to suit the signal, but in general the signal must to some extent, for reasons of economy, be tailored to suit the channel. One form of tailoring or 'signal processing' may thus be to pass the signal through a filter, in order to reduce and define its bandwidth prior to transmission. In multichannel telephony, for example, a 'commercial speech bandwidth' of 0·3 to 3·4 kHz has been internationally agreed. In systems where no filtering is arranged prior to transmission, the communication channel itself will effectively filter the signal by attenuating some components more than others. It is therefore important to study the distortion which arises when signals are transmitted through channels whose amplitude and phase characteristics, as functions of frequency, are not ideal. In particular, the effect of a low-pass filter upon an impulse, a voltage step, and a rectangular pulse will be studied. These problems have an important bearing upon the design of pulse communication systems and also in radar and in television. For idealised filters, these problems are relatively simple and will receive most attention. Certain general techniques, however, will be developed for finding the response of any linear system to an arbitrary input signal, in terms either of the amplitude and phase response of the system or of its impulse response.

The later sections of this chapter are devoted to the study of a quite different type of distortion which arises from non-linearity. The several links in a communication chain are usually designed to behave as linear networks. Non-linear distortion may, however, be put to useful service in frequency changing and in other types of signal processing such as modulation and detection. Frequency changing and also the effects of switching and sampling will be examined in the concluding sections of this chapter. The remaining and most vital signal processing techniques—modulation and demodulation—will be discussed in Chapter 4.

3.2 LINEAR SYSTEMS

Consider a two-port network which responds to an input signal $v_1(t)$ by giving the output signal $g_1(t)$. Similarly, $v_2(t)$ has as its response $g_2(t)$. Now let the input be

$$v(t) = av_1(t) + bv_2(t) \qquad 3.1$$

then the system is said to be *linear* if the output is

$$g(t) = ag_1(t) + bg_2(t) \qquad 3.2$$

Equations 3.1 and 3.2 express the principle of linear superposition. In the case of networks containing only passive L, C, and R elements, linear superposition is a direct consequence of the fact that the differential equations which govern the circuit are linear. This, in turn, derives from the linear proportionality of Ohm's law and the principle of linear superposition in electromagnetism.

Now let $v_1(t) = x(t)$ and $v_2(t) = x(t+\tau)$ so that $v_2(t)$ is identical to $v_1(t)$ but advanced by τ. Setting $a = -1/\tau$ and $b = 1/\tau$ in equation 3.1

$$v(t) = \frac{x(t+\tau) - x(t)}{\tau}$$

If $y(t)$ is the response to $x(t)$ then, by equation 3.2, the response to $v(t)$ is

$$g(t) = \frac{y(t+\tau) - y(\tau)}{\tau}$$

Allowing $\tau \to 0$, we obtain the important result for linear systems that if $y(t)$ is the response to $x(t)$ then $y'(t)$ is the response to $x'(t)$. Similarly, $y''(t)$ will be the response to an input $x''(t)$.

A remarkable property of linear systems is that a sinusoidal input signal gives rise, in the steady state, to a sinusoidal output signal. The amplitude and phase of the response may differ from that at the input, but the *shape* is unaltered. This is not true for other wave shapes (see, for example, Fig. 2.10). The rather special behaviour of sine waves in linear systems may be brought out from the principle of linear superposition as follows. Let $x(t)$ be some signal with derivatives $x'(t)$ and $x''(t)$. Writing $v_1(t) = x''(t)$ and $v_2(t) = x(t)$ with $a = 1$ and $b = \omega^2$ in equation 3.1

$$v(t) = x''(t) + \omega^2 x(t)$$

hence, from the result of the previous paragraph

$$g(t) = y''(t) + \omega^2 y(t)$$

For the particular case $x(t) = \cos \omega t$ the input $v(t)$ vanishes. The response to a null input must be zero, and hence $g(t) = 0$, i.e.

$$y''(t) + \omega^2 y(t) = 0$$

and so
$$y(t) = A \cos (\omega t + \psi) \qquad\qquad 3.3$$

i.e., the response of a linear system to a sinusoid is a sinusoid of the same frequency. Note that the input $x(t) = \cos \omega t$ exists for all t so that the above result relates to the *steady-state* response.

3.3 LINEAR SYSTEM RESPONSE BY FOURIER ANALYSIS

The values of A and ψ in equation 3.3 will in general depend on the frequency of the input sine wave. The steady-state response of a linear system may be characterised by the two functions

$$\left.\begin{array}{l} A(\omega) = \text{output amplitude/input amplitude} \\ \psi(\omega) = \text{output phase} - \text{input phase} \end{array}\right\} \qquad 3.4$$

For the *RC* integrating network of Fig. 3.1, for example

$$A(\omega) = \frac{1}{(1 + \omega^2 R^2 C^2)^{1/2}}$$

$$\psi(\omega) = -\tan^{-1}(\omega CR)$$

65

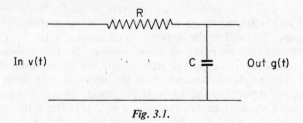

Fig. 3.1.

Now consider the input waveform whose Fourier expansion is

$$v(t) = C_0 + \sum_n C_n \cos (n\omega_1 t - \phi_n) \qquad 3.5$$

It should perhaps be stressed once more that equation 3.5 does not represent a waveform which is 'switched on' at $t = 0$. Each sinusoidal component has been present at the input ever since $t = -\infty$ and continues to $t = +\infty$. The transient response of the system to a sinusoid which begins at $t = 0$ is of no interest in this connection. The steady-state response to each input component is assumed to be known and the principle of linear superposition, equations 3.1 and 3.2, asserts that the output waveform will be given by

$$g(t) = A(0)C_0 + \sum_n A(n\omega_1)C_n \cos [n\omega_1 t - \phi_n + \psi(n\omega_1)] \qquad 3.6$$

In this way the (steady-state) response of the system to any periodic signal whose Fourier expansion is known, may be calculated from the functions $A(\omega)$ and $\psi(\omega)$.

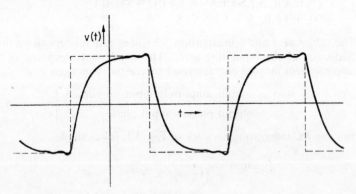

Fig. 3.2.

The method is illustrated in Fig. 3.2, in which the broken line shows a square wave at 159·2 Hz ($\omega_1 = 10^3$). This is applied to the circuit in Fig. 3.1 with $RC = 500$ μs. The functions $A(\omega)$ and $\psi(\omega)$ for this circuit were noted above, while the Fourier expansion of the input signal was found in equation 2.7 (here the d.c. term $V/2$ is omitted). For convenience we may set $V = \pi/2$ and write the input as

$$v(t) = \cos 10^3 t - 0\cdot333 \cos 3.10^3 t + 0\cdot200 \cos 5.10^3 t - \ldots$$

hence

$$g(t) = 0\cdot895 \cos (10^3 t - 0\cdot464) - 0\cdot185 \cos (3.10^3 t - 0\cdot983)$$
$$+ 0\cdot074 \cos (5.10^3 t - 1\cdot190) - \ldots$$

The sum of the first six terms in this series is sketched as the solid curve in Fig. 3.2, and shows a good degree of approximation to the exact result.

In this example the method is seen to be extremely laborious and the standard methods of circuit theory would have been easier and more appropriate. The method comes into its own, however, when the given information about the system consists of the functions $A(\omega)$ and $\psi(\omega)$ rather than the precise circuit details of some complicated network. For certain idealised systems and signals, the response may be found analytically, as will be shown below, and certain general conclusions may be drawn which apply approximately to real systems.

One general conclusion may already be drawn from a comparison of equations 3.5 and 3.6. If the system is such that

$$A(\omega) = \text{constant} = A \qquad 3.7$$

and

$$\psi(\omega) = -\omega t_d \qquad 3.8$$

where t_d is a constant, then the results of Section 2.6 show that the output waveform is a delayed replica of the input waveform, delayed by t_d and scaled by a factor A. The signal is not distorted, so that the above conditions on $A(\omega)$ and $\psi(\omega)$ represent the requirements for *distortionless transmission* by a linear system. If inversion of the signal waveform is not counted as distortion then the phase function may contain an added constant $\pm n\pi$. When n is odd, every Fourier component will effectively be inverted so that the whole signal becomes inverted. More generally, therefore

$$\psi(\omega) = -\omega t_d + n\pi \qquad 3.9$$

67

The conditions for distortionless transmission are sketched in Fig. 3.3.

It is convenient to classify distortion into *amplitude distortion (or attenuation distortion)* when $A(\omega)$ fails to satisfy equation 3.7, and *phase distortion* when $\psi(\omega)$ fails to satisfy equation 3.8. Although

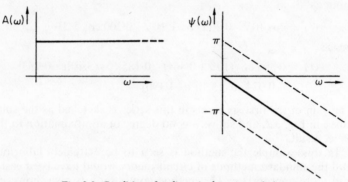

Fig. 3.3. Conditions for distortionless transmission

this distinction is convenient it should not be taken as implying that the functions $A(\omega)$ and $\psi(\omega)$ are independent. Indeed, it can be shown that it is impossible to design a physically realisable network which satisfies two independently specified amplitude and phase functions $A(\omega)$ and $\psi(\omega)$. Once $A(\omega)$ is specified, then the choice of $\psi(\omega)$ becomes restricted and vice versa. In practice both amplitude and phase distortion will occur simultaneously.

3.4 IMPULSE RESPONSE OF AN IDEALISED LOW-PASS FILTER (FOURIER SERIES METHOD)

Problems involving signal transmission through linear systems which attenuate strongly above a specified frequency are important in practice. To simplify the analysis, let us ignore for the moment the restriction just mentioned and represent such a system by an idealised low-pass filter whose gain is unity up to angular frequency ω_c and zero for $\omega > \omega_c$, while the phase shift is a linear function of frequency as shown in Fig. 3.4. As an input signal we will take the impulse train of Fig. 2.7(a) with the Fourier expansion

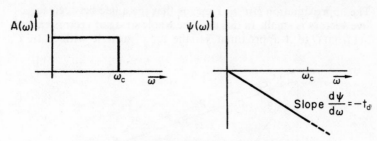

Fig. 3.4. Idealised low-pass filter characteristics

(equation 2.9),

$$v(t) = \frac{2k}{T}\left(\frac{1}{2} + \cos\omega_1 t + \cos 2\omega_1 t + \cos 3\omega_1 t + \dots\right)$$

The output will be

$$g(t) = \frac{2k}{T}\left[\frac{1}{2} + \cos\omega_1(t - t_d) + \cos 2\omega_1(t - t_a)\right.$$
$$\left. + \dots + \cos n\omega_1(t - t_d)\right]$$

Let $T = 2\pi/\omega_1$ be chosen to make ω_c/ω_1 take an integer value n. The expansion for $g(t)$ then terminates at the nth harmonic. Referring to a time origin at $t = t_d$, let $t' = t - t_d$, so that

$$g(t') = \frac{2k}{T}\left(\frac{1}{2} + \cos\omega_1 t' + \cos 2\omega_1 t' + \dots + \cos n\omega_1 t'\right) \quad 3.10$$

We are interested in the form of this function when ω_1 is small (T large) so that, for $|t'| \ll 1/\omega_1$, $g(t')$ describes the response to a single impulse at the input terminals.

For a given t', the sum expressed by equation 3.10 may be represented graphically as the projection onto the horizontal axis of the vector sum of $n + 1$ vectors. The first is of length k/T and all subsequent ones are of length $2k/T$, the angle of the $(r + 1)$th vector being $r\omega_1 t'$. This is illustrated in Fig. 3.5; the resultant vector has length R and angle θ, so that

$$g(t') = R\cos\theta$$

69

The approximation $\omega_1 t' \ll 1$ means that the angle between successive vectors is small. In this case the whole string of vectors extending from O to A approximates to the arc of a circle with centre C.

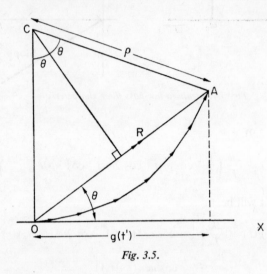

Fig. 3.5.

The arc length l and the angle θ are given by

$$l = 2(n+1/2)\,k/T$$
$$\theta = n\omega_1 t'/2$$

hence the radius ϱ is

$$\varrho = OC = CA = \frac{l}{2\theta}$$
$$= \frac{k(n+1/2)}{\pi n t'}$$

but, since ω_1 is small, $n = \omega_c/\omega_1$ is large, thus we may write approximately

$$\varrho = \frac{k}{\pi t'}$$

Now

$$R = 2\varrho \sin \theta$$

70

hence

$$g(t') = R \cos \theta$$
$$= \varrho \sin 2\theta$$
$$= \frac{k}{\pi t'} \sin n\omega_1 t'$$
$$= \frac{k}{\pi t'} \sin \omega_c t'$$
$$= \frac{k\omega_c}{\pi} \operatorname{sinc} \omega_c t'$$

and

$$g(t) = \frac{k\omega_c}{\pi} \operatorname{sinc} \omega_c(t - t_d) \qquad 3.11$$

This result is sketched in Fig. 3.6. Recalling the assumption $|\omega_1 t'| \ll 1$ (i.e., $|t'| \ll T$) it is seen that equation 3.11 is a satisfactory approximation for $|t'| \gg T$, becoming exact as $T \to \infty$. Thus *it describes the response of the filter to a single impulse of strength k at t = 0.* The output pulse is of amplitude $k\omega_c/\pi$, the central maximum being delayed by t_d, which is the slope of the

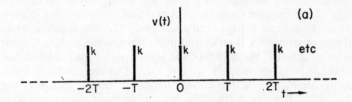

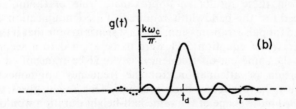

Fig. 3.6. (a) *Input train of impulses to low-pass filter of Fig. 3.4.* (b) *Response to the impulse at t = 0* ($|t'| \ll T$, see text)

71

phase characteristic of the filter. Before and after the principal maximum the response exhibits 'ringing' oscillations at the filter cut-off frequency. (There is a close analogy between the impulse response of a low-pass filter and the diffraction pattern produced by a plane wave at a single rectangular aperture. For a graphical treatment of Fraunhofer diffraction, similar to that given above, see for example Jenkins, F.A., and White, H.E., *Fundamentals of Optics* (McGraw-Hill, London, 1951).) The interval between the zeros on either side of the principal maximum of sinc $\omega_c t$ is given by $\Delta t = 2\pi/\omega_c = 1/f_c$. The width, τ, of the main peak at half-height turns out to be $\tau = 3 \cdot 79/\omega_c = 0 \cdot 603/f_c$ (see Appendix 1). As an approximate rule we may write $\tau \approx 1/2f_c$.

A somewhat disturbing feature of equation 3.11 is that $g(t)$ is finite for t negative, i.e. there is some response before the impulse is applied. Needless to say, this is impossible as it would defy the law of causality. The mistake lies in the initial assumptions, where the amplitude and phase characteristics of the filter were specified independently. As was noted earlier, this is not permissible and although filters having a sharp cut-off, approaching that of Fig. 3.4, may be constructed, their phase shift is not a linear function of frequency. The resulting phase distortion is always sufficient to ensure $g(t) \equiv 0$ for $t < 0$.

Equation 3.11 will often be a poor approximation to the detailed shape of the impulse response of a practical low-pass filter. Nevertheless, the general features remain, namely: (i) the response to an impulse of strength k is an output pulse whose amplitude is roughly $k\omega_c/\pi = 2kf_c$, (ii) the width of the output pulse measured at half amplitude is given roughly as $\tau = 1/2f_c$, (iii) the output exhibits ringing oscillations at the cut-off frequency before and after the main pulse. Ringing oscillations before the main pulse *(precursor* oscillations) may not be observed in practice unless the time delay t_d of the filter is fairly large.

From these results an approximate 'rule of thumb' may be derived for the bandwidth required of a communication link designed for pulse transmission. The first step is to note that if the *output* pulse $g(t)$ of equation 3.11 were to be applied to a second filter with the same cut-off frequency, it would be transmitted without distortion or attenuation, for the frequency components lying above f_c have already been removed. One might guess that any *rounded* pulse shape of the same half-height duration would also be transmitted satisfactorily, though perhaps with some distortion. A much shorter pulse could not be transmitted without severe distortion and attenuation since it would carry significant energy at frequencies well beyond the limit f_c. As shown above, if τ is the pulse duration at half-height

$$f_c \approx 1/2\tau \qquad\qquad 3.12$$

As a rough guide, therefore, if pulses of duration τ are to be transmitted, the system response must extend to at least $1/2\tau$ Hz. For example, a 1 µs pulse would require a bandwidth of 0·5 MHz. Referring to Fig. 2.16, it is seen that, for the pulse shapes dealt with there, the criterion $f_c\tau = 0·5$ of equation 3.12 means that at least 77% of the energy will be transmitted. The rectangular pulse, carrying significant components up to higher frequencies, will suffer most distortion and this case will be investigated more fully in Section 3.7 below.

3.5 STEP RESPONSE OF AN IDEALISED LOW-PASS FILTER

In Section 3.2 it was shown that, for a linear system, if $y(t)$ is the response to $x(t)$ then $y'(t)$ is the response to $x'(t)$. Conversely, the response to the integral of a certain signal will be the integral of the response to that signal, apart from an arbitrary constant. The integral of the impulse signal just considered is a step of height k and hence the response of a low-pass filter to a step function will be the integral of equation 3.11, namely

$$g(t) = \frac{V}{\pi}\left[Si\omega_c(t - t_d) + \frac{\pi}{2} \right] \qquad\qquad 3.13$$

in which the step height has been written V and the function $Si(x)$ has been defined earlier (equation 2.55, see also Appendix 1). Since $Si(x) \to -\pi/2$ for $x \to -\infty$, the choice of the arbitrary constant $V/2$ in equation 3.13 makes $g(t) \to 0$ for $t \to -\infty$. This input and response are sketched in Fig. 3.7. For a physically realisable filter, of course, $g(t) \equiv 0$ for $t < 0$.

The most significant feature of the response is the change from a sudden rise to a finite rise time. A convenient definition of rise time in this connection is that illustrated in Fig. 3.8. A tangent to the response is drawn at the steepest point, $t = t_d$, and the interval between the moments at which this line meets the asymptotes O

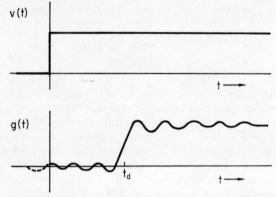

Fig. 3.7. *Input step function and response of idealised low-pass filter*

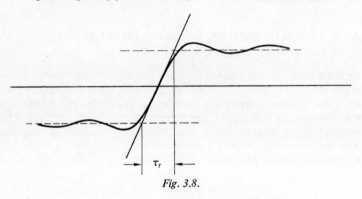

Fig. 3.8.

and V is taken as a measure of the rise time τ_r; thus, from equation 3.13

$$g'_{\max} = \frac{\mathrm{d}g}{\mathrm{d}t}\bigg|_{t=t_d} = \frac{\omega_c V}{\pi}$$

hence

$$\tau_r = \frac{V}{g'_{\max}}$$

$$= \frac{\pi}{\omega_c}$$

$$= \frac{1}{2f_c} \qquad\qquad 3.14$$

Once again, the filter considered here cannot be realised by any physical network. Nevertheless, the general features ot the response will be exhibited by practical filters and may be summarised as follows: (i) time delay—governed by the mean slope of the phase characteristic, (ii) 'overshoot' and 'undershoot' with oscillations at cut-off frequency f_c, (iii) a finite rise time given approximately by $\tau_r = 1/2f_c$. This last result gives another 'rule of thumb' for signal transmission: if the rise time τ_r of a wave front is to be preserved in transmission then the communication link must have an amplitude characteristic which remains flat up to a frequency $1/2\tau_r$ Hz or greater. A signal with a 10 ns rise requires a 50 MHz bandwidth, approximately.

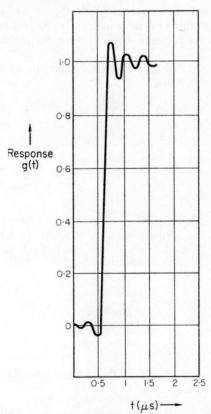

Fig. 3.9. Step response of a practical filter (from C. F. Floyd et al.[1])

The overshoot in the response is essentially the effect which was discussed in Section 2.3 as Gibbs's phenomenon. The $Si(x)$ function of equation 3.13 overshoots its asymptotic value by approximately 9% of the overall swing. The physical reality of the phenomenon is illustrated in Fig. 3.9, which shows the step response of a low-pass filter designed for television applications. (See reference 1. The design and step response of a number of low-pass filters are discussed in reference 2.)

3.6 APPLICATION OF THE FOURIER TRANSFORM

The results obtained in the previous section are more elegantly derived through the Fourier transform. It is first necessary, however, to define the function $H(\omega)$, the *transfer function* of a linear system.

In terms of the complex Fourier coefficients an arbitrary input sinusoid may be written

$$v(t) = a \cos(\omega t - \phi) = \alpha_\omega \, e^{j\omega t} + \alpha_{-\omega} \, e^{-j\omega t}$$

where

$$\alpha_\omega = \frac{a}{2} e^{-j\phi}$$

and

$$\alpha_{-\omega} = \frac{a}{2} e^{j\phi}$$

The output signal, with the notation of equations 3.4, is then

$$g(t) = aA(\omega) \cos[\omega t - \phi + \psi(\omega)]$$
$$= \beta_\omega \, e^{j\omega t} + \beta_{-\omega} \, e^{-j\omega t}$$

where

$$\beta_\omega = A(\omega) \, e^{j\psi(\omega)} \alpha_\omega \qquad \qquad 3.15$$

and

$$\beta_{-\omega} = A(\omega) \, e^{-j\psi(\omega)} \alpha_{-\omega} \qquad \qquad 3.16$$

Up to this point, $A(\omega)$ and $\psi(\omega)$ have been defined only for positive ω. If we now define

$$A(-\omega) = A(\omega) \quad \text{and} \quad \psi(-\omega) = -\psi(\omega)$$

and combine these into the complex function $H(\omega)$ where

$$H(\omega) = A(\omega)\,e^{j\psi(\omega)} \qquad 3.17$$

then equations 3.15 and 3.16 become

$$\beta_\omega = H(\omega)\,\alpha_\omega \qquad 3.18$$

Equation 3.18 applies for complex coefficients at both positive and negative frequencies, and expresses the relationship between the coefficients of the input and output waveforms. The relation between $H(\omega)$ and the measured amplitude and phase characteristics is sketched in Fig. 3.10 $\big(H(\omega)$ is identical to the Laplacian transfer function of the system with $p = j\omega.\big)$

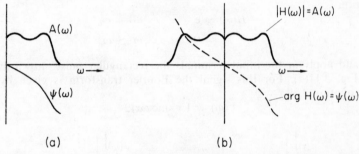

Fig. 3.10. (a) *Amplitude and phase characteristics related to* (b) *the complex transfer function* $H(\omega)$

In terms of the complex coefficients, the input and output waveforms of a periodic signal are now seen to be related as

$$v(t) = \sum_n \alpha_n\,e^{jn\omega_1 t} \qquad 3.19$$

$$g(t) = \sum_n \alpha_n H(n\omega_1)\,e^{jn\omega_1 t} \qquad 3.20$$

These equations are essentially the same as equations 3.5 and 3.6 and apply only to repetitive waveforms with fundamental frequency ω_1. From the derivation of the Fourier transform in Section 2.10, however, it is clear that for an input *pulse* $v(t)$ with output $g(t)$, equation 3.18 becomes

$$G(\omega) = H(\omega)\,V(\omega) \qquad 3.21$$

77

while 3.20 becomes

$$g(t) = \frac{1}{2\pi} \int_{-\infty}^{\infty} H(\omega)\, V(\omega) e^{j\omega t}\, d\omega \qquad\qquad 3.22$$

which is a general expression for the system response in terms of the transfer function $H(\omega)$ and the Fourier transform $V(\omega)$ of the input signal.

3.7 PULSE RESPONSE OF AN IDEALISED LOW-PASS FILTER

As an illustration of the Fourier transform method let us consider once again the idealised low-pass filter which has the transfer function

$$H(\omega) = e^{-j\omega t_d} \qquad |\omega| \leqslant \omega_c$$
$$= 0 \qquad\qquad |\omega| > \omega_c$$

and apply at the input terminals the rectangular pulse shown in Fig. 3.11(a). For this signal the Fourier transform is given by equation 2.39

$$V(\omega) = V\tau \,\text{sinc}\, \omega\tau/2$$

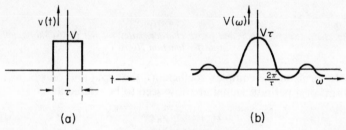

(a)   (b)

Fig. 3.11.

and is sketched in Fig. 3.11(b). Equation 3.21 now gives the Fourier transform of the output signal

$$G(\omega) = V\tau\, e^{-j\omega t_d} \,\text{sinc}\, \omega\tau/2 \qquad |\omega| \leqslant \omega_c$$
$$= 0 \qquad\qquad\qquad\qquad\quad |\omega| > \omega_c$$

The modulus of $G(\omega)$ is sketched in Fig. 3.12(a), which shows that filtering has simply removed components above ω_c, leaving the

78

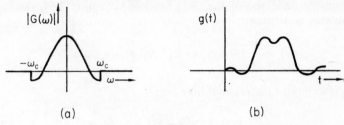

(a)   (b)

Fig. 3.12.

remainder of the spectrum unchanged apart from time delay.
The inverse transform of $G(\omega)$ given by equation 3.22 now reads

$$g(t) = \frac{1}{2\pi} \int_{-\infty}^{\infty} G(\omega)\, e^{j\omega t}\, d\omega$$

$$= \frac{V\tau}{2\pi} \int_{-\omega_c}^{\omega_c} \operatorname{sinc} \omega\tau/2\ e^{j\omega(t-t_d)}\, d\omega$$

or

$$g(t') = \frac{V\tau}{2\pi} \int_{-\omega_c}^{\omega_c} \operatorname{sinc} \omega\tau/2\ e^{j\omega t'}\, d\omega$$

where t' refers to a delayed time origin at $t = t_d$. Now, writing
$e^{j\omega t'} = \cos \omega t' - j \sin \omega t'$ and recalling that $\operatorname{sinc} \omega\tau/2$ is an even
function, it is seen that the imaginary part of the integral above
vanishes, leaving

$$g(t') = \frac{V\tau}{\pi} \int_0^{\omega_c} \frac{\sin \omega\tau/2 \cos \omega t'}{\omega\tau/2}\, d\omega$$

$$= \frac{V}{\pi} \left[\int_0^{\omega_c} \frac{\sin \omega(t'+\tau/2)}{\omega}\, d\omega - \int_0^{\omega_c} \frac{\sin \omega(t'-\tau/2)}{\omega}\, d\omega \right]$$

$$= \frac{V}{\pi} \left[\int_0^{\omega_c(t'+\tau/2)} \frac{\sin u}{u}\, du - \int_0^{\omega_c(t'-\tau/2)} \frac{\sin v}{v}\, dv \right]$$

with substitutions

$$u = \omega(t'+\tau/2)$$
$$v = \omega(t'-\tau/2)$$

79

From the definition of the $Si(x)$ function, equation 2.55, the last line may be written

$$g(t') = \frac{V}{\pi}[Si\,\omega_c(t'+\tau/2) - Si\omega_c(t'-\tau/2)]$$

or reverting to the original time origin

$$g(t) = \frac{V}{\pi}[Si\,\omega_c(t-t_d+\tau/2) - Si\,\omega_c(t-t_d-\tau/2)] \qquad 3.23$$

A typical form of this response is sketched in Fig. 3.12(b). This result might have been anticipated by regarding the rectangular input pulse as a step of height V at $t = -\tau/2$ followed by a second step $-V$ at $t = \tau/2$. The output is therefore a superposition of two expressions of the form of equation 3.13.

We are now in a position to examine the distortion of a rectangular pulse produced by filters of differing cut-off frequency. Fig. 3.13 shows the output pulse shapes corresponding to equation 3.23 when $f_c = \omega_c/2\pi$ is varied keeping τ fixed. When $f_c\tau \gg 1$ the output is a faithful replica of the input pulse, but distortion sets in as f_c is reduced. Setting $t = t_d$ in equation 3.23 yields an expression for the central amplitude of the response, namely

$$g_0 = \frac{2V}{\pi}Si\,\omega_c\tau/2 \qquad 3.24$$

which has a maximum value $g_0 = 1\cdot18\,V$ when $\omega_c\tau = 2\pi$, i.e. when $f_c = 1/\tau$. For values of f_c smaller than this the peak height of the response diminishes, falling to $g_0 = 0\cdot86\,V$ when $f_c = 1/2\tau$ and thereafter becoming rapidly smaller. For many applications involving pulse transmission (e.g., radar systems) the detailed shape of the received pulse is not important. It is sufficient to be able to detect the existence of the pulse, in which case a bandwidth $f_c = 1/2\tau$ represents a suitable lower limit, in agreement with the rough criterion developed in equation 3.12. Indeed, a bandwidth much larger than this will not improve the amplitude of the transmitted pulse significantly but will, as we shall see in Chapters 6 and 7, increase the amount of noise power received, possibly to such an extent as to mask the pulse completely. This means that the condition $f_c = 1/2\tau$ is close to the optimum as regards received signal-to-noise power ratio (see Chapter 7, Problem 5).

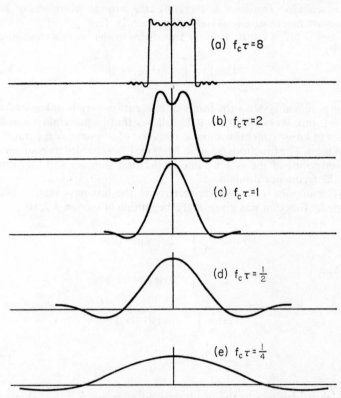

Fig. 3.13. *Response of idealised low-pass filter to a rectangular pulse of duration τ. Cut-off frequency varied, τ fixed*

3.8 IMPULSE RESPONSE OF A LINEAR SYSTEM

When a unit impulse at $t = 0$ is applied to a system with transfer function $H(\omega)$ we have (Section 2.12)

$$V(\omega) = 1$$

and hence from equation 3.22, the response $h(t)$ is

$$h(t) = \frac{1}{2\pi} \int_{-\infty}^{\infty} H(\omega) e^{j\omega t} \, d\omega \qquad 3.25$$

The impulse response is therefore the inverse transform of the transfer function. Conversely, the transfer function of a linear system is given by the Fourier transform of the impulse response

$$H(\omega) = \int_{-\infty}^{\infty} h(t)\, e^{-j\omega t}\, dt \qquad\qquad 3.26$$

For practical systems the limits of integration may be taken as 0 to $+\infty$, since $h(t) \equiv 0$ for $t < 0$. It follows that in principle it is sufficient to know *either* the impulse response of a system *or* the transfer function. The functions $h(t)$ and $H(\omega)$ may be regarded as equivalent descriptions of the system, one in the time domain and the other in the frequency domain.

The impulse response characterising the low-pass filter, whose transfer function was given at the beginning of section 3.7, is

$$
\begin{aligned}
h(t) &= \frac{1}{2\pi} \int_{-\omega_c}^{\omega_c} e^{-j\omega t_d}\, e^{j\omega t}\, d\omega \\
&= \frac{1}{2\pi} \int_{-\omega_c}^{\omega_c} \cos\omega(t - t_d)\, d\omega \\
&= \frac{1}{\pi} \int_{0}^{\omega_c} \cos\omega(t - t_d)\, d\omega \\
&= \frac{\omega_c}{\pi}\, \text{sinc}\,\omega_c(t - t_d) \qquad\qquad 3.27
\end{aligned}
$$

which is identical to the result derived after much more labour in equation 3.11. (The same result also follows from the limiting form of equation 3.23 when $\omega_c \tau \ll 1$, as may be seen in the lower curve of Fig. 3.13.)

3.9 RESOLUTION OF TWO IMPULSES. TELEVISION BANDWIDTH REQUIREMENTS

When two equal impulses separated by a time T are presented to a low-pass filter, the response will be the sum of two pulses having the form of equation 3.27. Fig. 3.14 shows the response for the cases $T = 1/2f_c$, $2/3f_c$ and $1/f_c$. It is seen that when T is too small it is not possible to decide from the shape of the output pulse whether

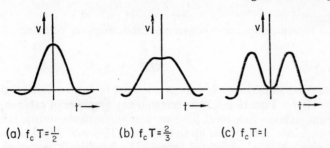

(a) $f_c T = \frac{1}{2}$ (b) $f_c T = \frac{2}{3}$ (c) $f_c T = 1$

Fig. 3.14. The response of a low-pass filter, cut-off frequency f_c, to two impulses separated by T

two impulses were applied at the input, or a single impulse of greater strength. As T is increased, a central dip appears in the response which grows deeper as T grows larger until the two pulses are clearly resolved. Taking the existence of a central minimum as a criterion for resolution of the two impulses by the system, we require approximately

$$T > 2/3f_c \qquad\qquad 3.28$$

There is a close analogy here with the optical resolving power of a telescope. The optical analogue of the electrical impulse is a point light source, such as a star. Due to diffraction effects the image is not a point but a diffraction pattern, and the ability of the instrument to resolve two stars whose angular separation is small is measured as the angular separation required for the composite diffraction image to exhibit a central minimum—this criterion is known as 'Rayleigh's criterion'.

Equation 3.28 is pertinent to the assessment of the ability of a television system to transmit fine detail in a picture. Suppose that a television camera is recording a scene which is totally dark except for two points of light which lie in the same plane as a scanning line. Assuming a perfect camera system (including perfect optics), the video signal will contain two electrical impulses separated by a time T which is determined by the frame repetition rate, the number of lines in the raster, and the distance between the points in the image. Let us suppose we will be satisfied if the transmission system allows the points to be resolved in the received picture when they are a distance apart equal to the line separation (this would be a little better than the definition of a typical broadcast picture). That is to say, the horizontal and vertical resolving powers are to be comparable.

For simplicity we may assume an aspect ratio of 1 : 1 and if there are N (complete) pictures per second each composed of L lines, then

$$T = \frac{1}{NL^2}$$

With $N = 25$ and $L = 625$ as in a typical broadcast system, then $T \approx 10^{-7}$ s. Equation 3.28 asserts that any amplifier or cable in the communication link must have an amplitude characteristic which remains substantially flat up to 6·7 MHz. This would be known as the video bandwidth of the signal. The bandwidth occupied by a modulated radio wave might exceed this, depending on the modulation method employed (see Chapter 4). The video bandwidth is inversely proportional to T and therefore directly proportional to the number of resolved 'points' in the picture. The greater the picture detail, and thus the greater the rate of information transmission, the wider the required bandwidth—illustrating once again the general rule of Section 2.14.

3.10 PHASE DISTORTION. GROUP DELAY AND PHASE DELAY

Up to this point our attention has been directed entirely towards amplitude distortion. It is easily seen that amplitude distortion is always symmetrical in its effect. Consider a symmetrical pulse applied to a system whose phase characteristic is linear, so that there is no phase distortion. The Fourier transform of the pulse has even symmetry, as must $A(\omega)$ by definition. It follows that, apart from the delay factor $e^{-j\omega t_d}$, the transform of the output pulse is also real and even and thus the output pulse is symmetrical about the delayed time origin $t = t_d$. Phase distortion, on the other hand, always gives rise to asymmetry, as may be seen from the following example.

Consider a fictitious system whose phase characteristic is that of Fig. 3.15. For simplicity the phase lag at high frequencies will be taken as 2π and the gradient of the initial characteristic taken as

$$-\frac{d\psi}{d\omega} = t_d = \frac{\pi}{2\omega_1} = \frac{1}{4f_1}$$

The amplitude characteristic is assumed to be unity for all ω. As an input pulse let us take a rectangular pulse of duration τ, where

$$\tau = t_d = \frac{1}{4f_1}$$

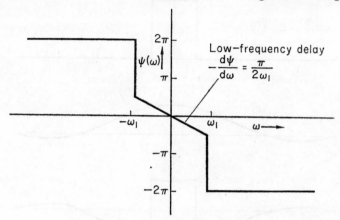

Fig. 3.15. Phase response of a hypothetical filter

Reference to Fig. 3.13(e) shows that the low-frequency components of this pulse, i.e. those for which $|\omega| < \omega_1$, combine to give the pulse shown in Fig. 3.16(b) and subtracting this from the original pulse, Fig. 3.16(a), reveals the contribution from the high-frequency components, $|\omega| > \omega_1$, to be that of Fig. 3.16(c). The rectangular pulse applied has now been split into two 'component pulses' whose spectra are confined to the frequencies below and above ω_1, respectively. The transmission of these may be considered separately. The 'low-frequency component' (b) will be transmitted undistorted but with delay t_d to become that shown at (d). The 'high-frequency component', (c), will be transmitted undistorted and without delay, since the phase change (2π) is constant for all the frequencies in this pulse. By superposition, the output pulse is that of the bottom curve in Fig. 3.17. The existence of a response before the input pulse arrives shows that, once again, the system transfer function envisaged could not be realised in practice. Nevertheless, this example suffices to show (i) that severe phase distortion can give rise to ringing oscillations in the response and (ii) that phase distortion is asymmetric in its effect. When amplitude and phase distortion are present together in a real system, the asymmetry is always such as to ensure zero response before the input pulse arrives.

A somewhat different approach to phase distortion is appropriate when considering transmission over a continuous structure such as a cable. The phase lag per unit length of an unloaded cable commonly has the form shown in Fig. 3.17. Suppose that a pulse of

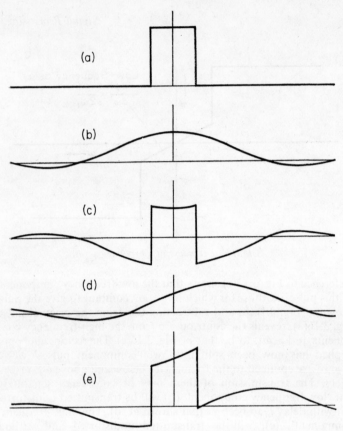

Fig. 3.16. (a) *Input pulse to system of Fig. 3.15* (b) *Low-frequency components* (c) *High-frequency components* (d) *Delayed version of* (b) (e) *Output pulse formed as* (c) + (d)

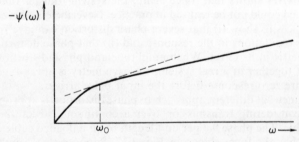

Fig. 3.17.

oscillations $\cos \omega_0 t$ lasting for a time τ is applied at one end of such a cable. This pulse and its spectrum were discussed in Chapter 2 as transform I of Fig. 2.15. Almost the whole energy of the pulse is confined to a narrow region in the frequency domain around $\omega = \omega_0$, so that the phase characteristic may be approximated as linear, with a slope given by the tangent at $\omega = \omega_0$ (see Fig. 3.17). The 'wave packet' will travel along the cable, arriving at a distance L from the sending end after a time

$$t_d = \frac{\mathrm{d}\psi}{\mathrm{d}\omega}\bigg|_{\omega_0}$$

where ψ is the phase lag associated with a length L of the cable. If λ is the wavelength at angular frequency ω_0, then

$$\psi = \frac{2\pi L}{\lambda}$$

hence

$$t_d = 2\pi L \frac{\mathrm{d}(1/\lambda)}{\mathrm{d}\omega}$$

The velocity of propagation of the wave packet is thus

$$v_g = \frac{L}{t_d} = \frac{1}{2\pi}\frac{\mathrm{d}\omega}{\mathrm{d}(1/\lambda)}$$

which is the usual expression for the *group velocity* of a wave in a dispersive medium. The delay associated with the envelope of a wave packet which comprises a group of frequencies close to ω is given by $t_d = \mathrm{d}\psi/\mathrm{d}\omega$ and is called the *group delay*. On the other hand, the phase velocity of the wave is related to the *phase delay* defined as ψ/ω. The transmission of some arbitrary signal may be considered by dividing its spectrum into small 'groups', and unless these arrive simultaneously at the receiving end distortion will occur. Absence of phase distortion thus requires $t_d = \mathrm{d}\psi/\mathrm{d}\omega = \text{con-}$ stant as we have seen before, and for this reason phase distortion is sometimes called *delay distortion*.

When signals are to be transmitted over long cables it is necessary to insert amplifiers (repeaters) at intervals in order to overcome the attenuation of the cable. The attenuation is frequency-dependent so that equalising networks are also employed in order to maintain a constant overall amplitude response $A(\omega)$ within the working bandwidth. This corrects for amplitude distortion and in telephone circuits this is often sufficient, as the ear is not very sensitive to phase distortion. (For a discussion on phase distortion in telephony see Steinberg.[3]) With lumped-loaded telephone cables, however, the phase characteristic curves sharply upward at the upper end of the transmission band. In a long circuit this may be sufficient to cause noticeable squeals at the ends of spoken syllables, due to the larger group delay at higher frequencies (see Guillemin,[4] p. 494). This distortion may be overcome by adding, at each repeater section, a delay-equalising network, designed to provide an overall linear phase characteristic. For the transmission of television signals, faithful reproduction of the wave shape is more important than in telephony and careful attention is paid to both attenuation and delay-equalising networks in long cable circuits. Some details of a practical system may be found in references 5 and 6 at end of this chapter.

3.11 CONVOLUTION

Thus far the response of a linear system has been related to the system transfer function by calculations performed in the frequency domain—equations 3.21 and 3.22. Equation 3.22 may be transformed into a time domain integral using equation 2.43, thus

$$g(t) = \frac{1}{2\pi} \int_{-\infty}^{\infty} H(\omega)\, V(\omega)\, e^{j\omega t}\, d\omega$$

$$= \int_{-\infty}^{\infty} h(\tau)\, v(t-\tau)\, d\tau$$

$$= \int_{-\infty}^{\infty} h(t-\tau)\, v(\tau)\, d\tau$$

where $h(\tau)$ is the response of the system to a unit impulse (see Section 3.8). For real systems $h(\tau) \equiv 0$ for $\tau < 0$, and if we also consider the input signal $v(t)$ to be applied at $t = 0$ so that $v(\tau) \equiv 0$ for $\tau < 0$, then the limits in the last two integrals may be restricted to the range 0 to t

$$g(t) = \int_0^t h(\tau)\, v(t-\tau)\, d\tau = \int_0^t h(t-\tau)\, v(\tau)\, d\tau \qquad 3.29$$

The first of these equations has a simple physical interpretation.
Consider the input signal $v(t)$ not as a smooth waveform but as
a series of adjacent rectangular pulses of duration $d\tau$. The response
at time t is the sum of the responses to these impulses. The input
impulse at $t-\tau$ had a strength $v(t-\tau)\,d\tau$ and since this occurred τ
seconds ago its contribution to $g(t)$ is given as $h(\tau)\,v(t-\tau)\,d\tau$. The
first of the equations 3.29 adds all these contributions together.
Note that the use of superposition again implies the system is
linear.

A useful graphical aid to understanding the convolution integral
may be obtained from the second of equations 3.29. This is illustrated in Fig. 3.18. The upper trace (a) represents an impulse response

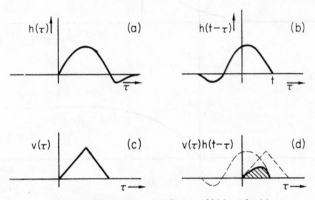

Fig. 3.18. Illustrating convolution of $h(\tau)$ with $v(\tau)$

$h(\tau)$. When this is reflected about $\tau = 0$ and then slid along the
τ-axis to the point $\tau = t$ we have the function $h(t-\tau)$ shown as
trace (b). Trace (c) represents an input signal $v(\tau)$. The product
$v(\tau)\,h(t-\tau)$ is shown in (d) and the integral under this curve represents $g(t)$, i.e. $g(t)$ is the shaded area in (d). A simple practical example is shown in Fig. 3.19. An integrating RC circuit and its impulse
response are sketched in (a). The convolution process is illustrated
in (b) for an input voltage step. It is readily seen that the area under
the product of the curves in (b), as the 'reversed impulse function'
slides to the right, gives the well-known step response of the RC
circuit shown at (c). The reader is invited to imagine the convolution of a step input function with the sinc $\omega_c t$ impulse response
of an ideal low-pass filter (Fig. 3.6, and compare with Fig. 3.7).

7*

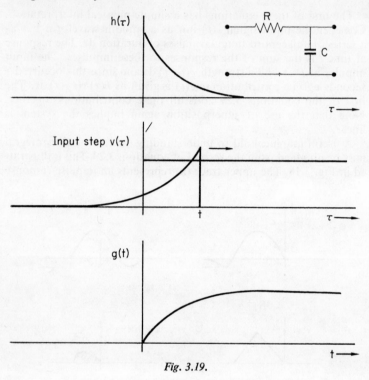

Fig. 3.19.

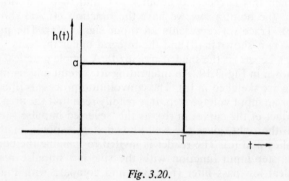

Fig. 3.20.

The time-domain analogue of the low-pass filter is a hypothetical filter having the rectangular impulse response shown in Fig. 3.20

$$h(t) = a \qquad 0 < t \leqslant T$$
$$h(t) = 0 \qquad t \leqslant 0 \quad \text{and} \quad t > T$$

The response to a signal $v(t)$ is easily found from equation 3.29, or the graphical method, to be

$$g(t) = a \int_{t-T}^{t} v(\tau) \, d\tau \qquad\qquad 3.30$$

For the case $a = 1/T$ the output $g(t)$ is seen to be the *average* value of the input signal over the previous T seconds; this filter is a *time-averaging filter*. As a hypothetical device it can sometimes be useful in signal theory—see for example Section 6.4. In practice the arrangement of Problem 6 provides a close approximation to this filter.

3.12 TESTING LINEAR SYSTEMS

We have seen that a linear system may be specified in terms of its transfer function $H(\omega)$ and it is common to test a signal transmission system by measuring its amplitude and phase characteristic over the frequency range of interest. It has been shown in Sections 3.8 and 3.11 above, however, that a measurement of the impulse response $h(t)$ also completely characterises the system.

When the impulse response exhibits a well-developed principal maximum whose width at half-height is T then it is easily seen by convolution that the response to a sinusoidal input signal at frequency f will be small when $fT \gg 1$, for the convolution integral then extends over many alternately positive and negative half-cycles. As a rough criterion it may be argued that the steady-state sine wave response will decline at frequencies above that for which T corresponds to one half-cycle, i.e. that the system attenuates strongly above a frequency f_h where

$$f_h \approx \frac{1}{2T} \qquad\qquad 3.31$$

Examination of the impulse response can in this way give a very rapid estimate of the upper frequency limit of the system. The same information can also be obtained from the response to a voltage step. By convolution it is easily seen that the rise time of the

91

response to a step is approximately equal to T so that f_h may also be estimated as the reciprocal of twice the rise time (cf. equation 3.14).

In a practical observation of the impulse response, the mathematical input impulse must be approximated by using a finite pulse whose duration must be much shorter than T. Care must be taken to ensure that the input pulse height is not so great as to drive the system into a non-linear region and this may mean that the strength of the testing impulse is so restricted that the response is inconveniently small. From this point of view an input step function may be preferred.

Both the impulse and the step function, however, have spectral components extending well beyond the upper frequency limit of the system. When testing communication links designed to carry component frequencies which have a well-defined upper limit, a testing pulse having a more restricted spectrum may give a clearer indication of the behaviour of the system *within* the transmission band. Such a test pulse is the 'cosine-squared' pulse of Fig. 2.14 E. Cosine-squared pulses have been used in the testing of television links and, with f_h as the upper frequency limit of the link, a pulse of this shape with width at half-height given by $T = 1/2f_h$ is known as a 'cosine-squared T-pulse'. The choice of pulse width is governed by the fact that $1/2f_h$ is roughly the width at half-height of the impulse response (equation 3.31) and thus measures the duration of the smallest picture element (a black or white dot) which can be transmitted. The spectrum of the T-pulse (see Fig. 2.14) falls to zero at $f = 1/T = 2f_h$ and is negligible beyond this frequency. This is adequate to test the system right across the upper edge of the transmission band and empirical limits may be set to the permitted amount of distortion in the received pulse for the system to be classed as acceptable. Further details on this method of testing may be found in references 7 and 8 at the end of this chapter.

3.13 NON-LINEAR DISTORTION

Having dealt at some length with the techniques for describing and analysing the behaviour of linear systems, we now turn to examine some of the characteristics of the response of non-linear systems. The circuit of Fig. 3.21(a) will serve as a simple model to begin the discussion. With the input terminals short-circuited, the current flowing, and hence v_0, is determined by E_b, R, and the diode characteristic. If an e.m.f. v_i is now connected across the input terminals, v_0 will change in a non-linear fashion, as shown in Fig. 3.21(b).

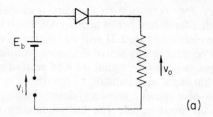

(a)

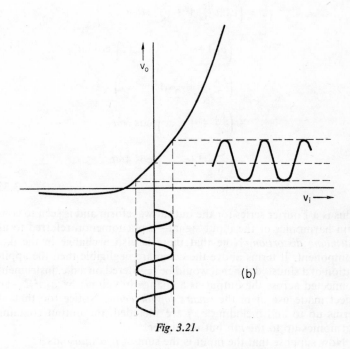

(b)

Fig. 3.21.

The characteristic may be expanded as a power series

$$v_0 = c_0 + c_1 v_i + c_2 v_i^2 + c_3 v_i^3 + \ldots \qquad 3.32$$

For small input signals the quadratic and higher terms may be neglected and the circuit is essentially linear. As v_i is made larger, terms of higher order must be included, depending on the values of the coefficients c_0, c_1, $c_2 \ldots$. (For a given maximum signal excursion

93

a satisfactory approximation to the curved characteristic may be obtained with a finite number of terms.) Although in the discussion to follow the circuit of Fig. 2.21 will be taken as a model, the arguments apply equally to any active device or complete amplifier for which the change in output signal may be related to the change in input signal through an expansion of the above form.

Fig. 2.21(b) also shows a graphical determination of the response to a sinusoidal input signal. The output is seen to be distorted. Setting $v_i = A \cos \omega t$ and inserting this into equation 3.32, we obtain

$$v_0 = \left(c_0 + \frac{c_2}{2} A^2 + \frac{3c_4}{8} A^4 + \ldots \right)$$

$$+ \left(c_1 A + \frac{3c_3}{4} A^3 + \ldots \right) \cos \omega t$$

$$+ \left(\frac{c_2}{2} A^2 + \frac{c_4}{2} A^4 + \ldots \right) \cos 2\omega t$$

$$+ \left(\frac{c_3}{4} A^3 + \ldots \right) \cos 3\omega t$$

$$+ \left(\frac{c_4}{8} A^4 + \ldots \right) \cos 4\omega t$$

$$+ \text{etc.} \qquad\qquad 3.33$$

This is a Fourier series for the output waveform and is seen to contain harmonics of the input signal—a phenomenon referred to as *harmonic distortion*. Note that there is also a change in the d.c. component. If terms above the cubic are negligible then the application of a sinusoidal signal would be registered on a d.c. instrument connected across the output as a change in voltage by $c_2 A^2 / 2$—an effect made use of in the *square law detector*. Notice too that, if terms up to and including $c_n v_i^n$ are included, the output contains harmonics up to the nth but no higher.

Now suppose that the input is the sum of *two* sinusoids

$$v_i = A \cos \omega_A t + B \cos \omega_B t$$

Substituting this into equation 3.32 and examining the result term by term, it is seen that the linear term gives output components at ω_A and ω_B, the quadratic term yields $\cos^2 \omega_A t$ and $\cos^2 \omega_B t$ and thus components at $2\omega_A$ and $2\omega_B$—i.e. harmonic distortion is present on each input component. The quadratic term also yields the cross-product

$$2c_2AB \cos \omega_A t \cos \omega_B t = c_2AB[\cos (\omega_A - \omega_B)t + \cos (\omega_A + \omega_B)t]$$

and thus the output contains components at $\omega_A \pm \omega_B$. Similarly, the cubic term gives components at $2\omega_A \pm \omega_B$ and $2\omega_B \pm \omega_A$ as well as at $3\omega_A$ and $3\omega_B$. Proceeding in this way we see that the output waveform contains sinusoidal components at the harmonics of the input signal frequencies together with terms at the frequencies $l\omega_A \pm m\omega_B$ where l and m are integers. These latter terms are called *intermodulation terms*. In the transmission or reproduction of music, ω_A and ω_B might be the fundamental components of the sounds produced by two instruments playing together. The generation of harmonics at either frequency is not necessarily unpleasant, since these are in any case produced by the musical instruments themselves. The intermodulation frequencies, on the other hand, are not present in the original sound wave (unless ω_B happens to be a precise harmonic of ω_A) and moreover $l\omega_A \pm m\omega_B$ need not be harmonically related in any pleasing way to either ω_A or ω_B. This musically unpleasant result is termed *intermodulation distortion*, though the non-linearity which is its basic cause means that harmonic distortion is also present simultaneously.

If the coefficients beyond c_n in equation 3.32 are all zero, the intermodulation frequencies $l\omega_A \pm m\omega_B$ are restricted to values of l and m such that $l+m \leqslant n$. The table below shows the amplitudes of the harmonics and intermodulation terms when terms up to the cubic are included.

ω	*Amplitude*	ω	*Amplitude*
d.c.	$c_0 + \frac{1}{2}c_2(A^2 + B^2)$	$3\omega_A$	$\frac{1}{4}c_3A^3$
ω_A	$c_1A + \frac{3}{4}c_3A(A^2 + 2B^2)$	$3\omega_B$	$\frac{1}{4}c_3B^3$
ω_B	$c_1B + \frac{3}{4}c_3B(B^2 + 2A^2)$	$\omega_A \pm \omega_B$	c_2AB
$2\omega_A$	$\frac{1}{2}c_2A^2$	$2\omega_A \pm \omega_B$	$\frac{3}{4}c_3A^2B$
$2\omega_B$	$\frac{1}{2}c_2B^2$	$2\omega_B \pm \omega_A$	$\frac{3}{4}c_3AB^2$

Signals and Information

3.14 FREQUENCY CONVERSION I

When the input signal to a non-linear device contains more than two sine waves the situation becomes much more complex and the output will contain components (i) at the frequencies of the input components, (ii) at their harmonics, and (iii) at the sums and differences of all frequencies in (i) and (ii). There is, however, a useful and interesting situation in which the output spectrum becomes much more simply related to the input spectrum. This occurs when one of the input components has a much larger amplitude than any of the others. Consider the input signal

$$v_i = X \cos \omega_X t + A \cos \omega_A t + B \cos \omega_B t$$

in which $X \gg A$ and $X \gg B$. When inserted into equation 3.32 the term $c_n v_i^n$ yields

$$c_n(X \cos \omega_X t + A \cos \omega_A t + B \cos \omega_B t)^n$$

$$= c_n X^n \left[\cos^n \omega_X t + \frac{nA}{X} \cos^{n-1} \omega_X t \cos \omega_A t + \frac{nB}{X} \cos^{n-1} \omega_X t \cos \omega_B t \right.$$

$$\left. + \frac{n(n-1)AB}{X^2} \cos^{n-2} \omega_X t \cos \omega_A t \cos \omega_B t + \dots \right] \quad 3.34$$

Now $\cos^n \omega_X t$ and $\cos^{n-1} \omega_X t$ may be expanded as trigonometric series in $\cos \omega_X t$, $\cos 2\omega_X t$, $\cos 3\omega_X t$, etc. It is then seen that the first term on the right of this equation gives rise to components at the harmonics of ω_X. The second and third terms give rise to products of the form

$$\cos l\omega_X t \cos \omega_A t = \tfrac{1}{2} \cos (l\omega_X + \omega_A)t + \tfrac{1}{2} \cos (l\omega_X - \omega_A)t \quad 3.35$$

If X is sufficiently large, the fourth and all remaining terms may be neglected. This means that all other intermodulation terms (such as for example $\omega_X \pm \omega_A \pm \omega_B$ or $2\omega_X \pm \omega_A \pm \omega_B$) are negligible as are also the harmonics of ω_A and ω_B. The only significant terms in the output are now (i) ω_X and its harmonics and (ii) terms differing from ω_X and its harmonics by $\pm \omega_A$ and $\pm \omega_B$. Furthermore, equation 3.34 shows that this latter group of components has amplitudes which are *linearly proportional* to the amplitudes of the corresponding input components $A \cos \omega_A t$ and $B \cos \omega_B t$. The argument may be extended to include further terms in the input signal, with amplitudes small compared with X, and each will give rise to sum and difference terms about the harmonics of ω_X which are linearly proportional to the corresponding input components.

This result is illustrated in Fig. 3.22, which shows at (a) an incoming signal with components lying in the frequency range ω_1 to ω_2 to which is added a strong locally generated sinusoid (or *heterodyne* signal) $X \cos \omega_X t$. This combination is applied to a

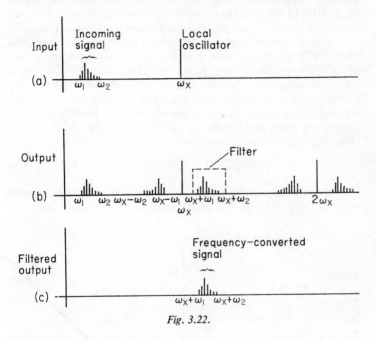

Fig. 3.22.

non-linear device which will provide an output signal whose spectrum (b) contains the terms discussed above. If the output is applied to a band-pass filter whose pass band extends from $\omega_X + \omega_1$ to $\omega_X + \omega_2$ then the filtered output shown at (c) is seen to have the spectrum of the incoming signal (a), but with each component increased in frequency by ω_X. Moreover, each component in this *frequency-converted* signal has an amplitude which is linearly proportional to the corresponding component in the incoming signal. The process may be reversed by taking the signal (c), adding to it a strong sinusoid $X \cos \omega_X t$, applying the combination

97

to a non-linear circuit, and passing the output through a filter with pass band ω_1 to ω_2. The output would then be an undistorted replica of the original signal at (a). Thus the combined processes of non-linear mixing and filtering provide what is essentially a linear operation; a signal may be shifted up and down the frequency axis at will and, on being returned to its original position, will have suffered no distortion. It may sometimes be preferable to select the difference frequencies lying between $\omega_X - \omega_2$ and $\omega_X - \omega_1$ in Fig. 3.22(b). In this case the order of the components is temporarily reversed, a matter of no great consequence in the usual applications. It is extremely useful to be able to move signals about like this in the frequency domain in order either to utilise some vacant part of the spectrum or else to suit the characteristics of a particular communication channel. Many methods of implementing this operation exist, besides the somewhat idealised one considered here. Before considering them, however, it will be found instructive to analyse a particular numerical example in greater detail.

Fig. 3.23 shows the response curve

$$v_0 = c_0 + c_1 v_i + c_2 v_i^2 + \ldots + c_8 v_i^8 \qquad 3.36$$

in which the values of the coefficients listed in the figure have been chosen to approximate the characteristics of an ideal diode for

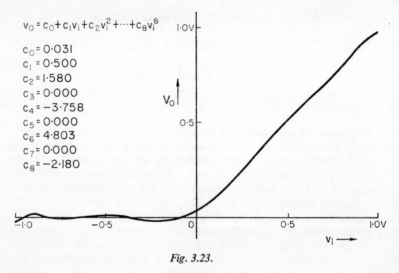

$v_0 = c_0 + c_1 v_i + c_2 v_i^2 + \cdots + c_8 v_i^8$

$c_0 = 0.031$
$c_1 = 0.500$
$c_2 = 1.580$
$c_3 = 0.000$
$c_4 = -3.758$
$c_5 = 0.000$
$c_6 = 4.803$
$c_7 = 0.000$
$c_8 = -2.180$

Fig. 3.23.

which

$$v_0 = 0 \qquad v_i < 0$$
$$v_0 = v_i \qquad 0 < v_i < 1$$

With this diode connected in the circuit of Fig. 3.21(a), (and with $E_0 = 0$) we apply an input signal

$$v_i = X \cos \omega_X t + A \cos \omega_A t$$

Inserting this expression into equation 3.36 and picking out the term at the difference frequency $\omega_X - \omega_A$ shows that v_0 contains the frequency-converted components given (with $X \gg A$) by

$$A\left[c_2 X + \frac{3c_4}{2} X^3 + \frac{15}{8} c_6 X^5 + \frac{35}{16} c_8 X^7 \right] \cos(\omega_X - \omega_A)t$$

As expected, the amplitude of this term is linearly proportional to A. Its amplitude relative to A is termed the *conversion efficiency* of the frequency converter and for this particular example the conversion efficiency is given by

$$\eta = c_2 X + \frac{3c_4}{2} X^3 + \frac{15}{8} c_6 X^5 + \frac{35}{16} c_8 X^7 \qquad 3.37$$

Notice that η is a function of the local oscillator amplitude X

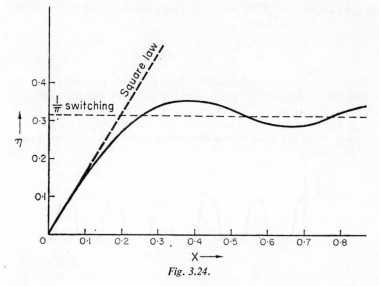

Fig. 3.24.

99

(always keeping $X \gg A$). With the coefficients of Fig. 3.23, the variation of η with X is shown in Fig. 3.24. It is seen that when X is small the conversion efficiency is proportional to X and is controlled by the quadratic coefficient c_2. For this reason, operation under these conditions is known as *square law frequency conversion* and is similar to the square law detector action mentioned in the previous section. More efficient frequency conversion is obtained by increasing the amplitude of the local oscillator drive, but it is interesting to note that the curve quickly levels off. If the ideal switching action had been approximated more closely, by including further terms in equation 3.36, the curve would have levelled off to a conversion efficiency of $1/\pi$ (see below). The example chosen, however, suffices to show that there will be a transition from square law operation to a 'switching mode' as X is increased. The diode is then acting as a switch whose opening and closing is controlled entirely by the local oscillator signal. Once this mode of operation is reached there is little point in increasing X still further.

3.15 FREQUENCY CONVERSION II

A different approach to the analysis of frequency conversion makes use immediately of the condition $X \gg A$. The local oscillator is regarded as controlling the small-signal conductance of the circuit, given for Fig. 3.21 as

$$g = \frac{\mathrm{d}I}{\mathrm{d}v_i} = \frac{1}{R} \frac{\mathrm{d}v_0}{\mathrm{d}v_i}$$

With $v_i = X \cos \omega_X t$ the conductance g varies periodically at the heterodyne frequency as the working point moves up and down the curve v_0 versus v_i. From the form of the characteristic in Fig. 3.21 the function $g(t)$ could be deduced and might be as shown in Fig. 3.25, the form being dependent upon the value of X. The

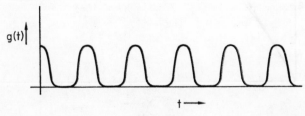

Fig. 3.25.

incoming signal $A \cos \omega_A t$ is now regarded as a small perturbation so that the *additional* current flowing due to the signal at frequency ω_A is given by

$$\Delta i = (A \cos \omega_A t) \times g(t)$$

From the form of the non-linear characteristic and for a given X a Fourier expansion of $g(t)$ may be made, giving

$$\Delta i = (A \cos \omega_A t)(g_0 + g_1 \cos \omega_X t + g_2 \cos 2\omega_X t + \ldots) \quad 3.38$$

which contains a component at the difference frequency

$$i_- = \frac{A g_1}{2} \cos(\omega_X - \omega_A)t \quad 3.39$$

and thus

$$\eta = \frac{g_1 R}{2} \quad 3.40$$

For the ideal switching diode $g(t)$ takes the form of a square wave, the conductance of the circuit switching from 0 to $1/R$, thus

$$g(t) = \frac{1}{2R} + \frac{2}{\pi R}(\cos \omega_X t - \cos 3\omega_X t + \ldots)$$

(cf. equation 2.7). Inserting $g_1 = 2/\pi R$ into equation 3.30 yields

$$\eta = \frac{1}{\pi}$$

which agrees satisfactorily with the approximate treatment of the switching mode given in the previous section.

A convenient way of selecting the desired output component at frequency $\omega_X - \omega_A$ (or $\omega_X + \omega_A$) is to use a tuned circuit instead of the load resistor. Fig. 3.26 shows the practical circuit of a simple fre-

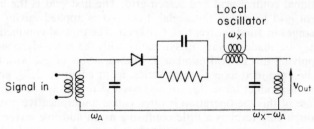

Fig. 3.26.

101

quency changer. Only the components of current close to the differ-ence frequency develop an output voltage, so that the load itself acts as a filter. The analysis of this circuit is somewhat more compli-cated than that given above, however, since the circuit conduct-ance is now frequency-dependent. It may be shown that with a resonant load, and taking into account the exponential form of practical diode characteristics, the conversion efficiency of this type of circuit approaches 1 rather than the value $1/\pi$ obtained above. Detailed analysis of this and other types of frequency changer may be found in the literature (e.g., references 9, 10 and 11). Instead of a diode, a transistor may be used, as in Fig. 3.27.

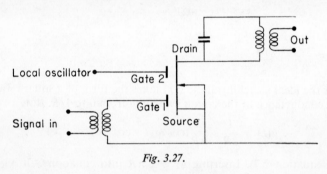

Fig. 3.27.

The type of frequency converter considered above operates by *adding* the incoming signal to the local oscillator drive and applying the combination to a non-linear device. This kind of operation is sometimes called 'additive mixing'. There are other types of fre-quency converter which use multigrid valves or special transistors in which the signal and local oscillator are applied to different terminals of the mixing device. One device in this class is the penta-grid frequency converter which may be viewed as a pentode with an additional control grid and screen grid. The first grid is the usual control grid to which the signal $A \cos \omega_A t$ is applied, giving rise to changes in anode current $g_m A \cos \omega_A t$. The mutual conductance g_m may be made to vary in sympathy with the heterodyne signal by applying this to grid number 3. The variation of g_m with time may be expanded as a Fourier series, as in equation 3.38, giving rise to a product term $A g_1 \cos \omega_X t \cos \omega_A t$ in the anode current. Because of this the operation is often called 'multiplicative mixing' (although this term is a little confusing as an additive mixer also works by generating the same product term). A resonant circuit

102

anode load is again used to select the desired sum or difference frequency $\omega_X \pm \omega_A$.

The more recently developed dual-gate m.o.s. field effect transistor works on the same principle. Here the drain current is controlled by the potentials applied to both gates 1 and 2. Usually, as in Fig. 3.27, the local oscillator is applied to gate No. 2 and this causes the mutual conductance between gate 1 and drain to fluctuate at heterodyne frequency. The incoming signal is applied to gate 1 and the sum or difference frequency is selected by a resonant load in the drain circuit. Practical details may be found in reference 12.

The approach used in this section shows that both additive and multiplicative mixing may be analysed in terms of the linear equations used for small-signal analysis, but in these equations the coefficients (e.g., the small-signal conductance or mutual conductance) are time-varying. The time variation of the coefficients is controlled by the local oscillator drive, either through non-linearity in the mixing device or else by way of a separate control electrode. As far as the small incoming signal is concerned, the system is instantaneously linear, with the result noted earlier—that frequency conversion is essentially a linear operation.

3.16 SWITCHING AND SAMPLING

It has been seen above that an extreme form of frequency converter effectively switches the signal on and off at the local oscillator frequency. An efficient way of accomplishing this is by use of the bridge circuit of Fig. 3.28. The diodes are arranged so that when the local drive, $X \cos \omega_X t$, is on one half-cycle the diodes are turned off.

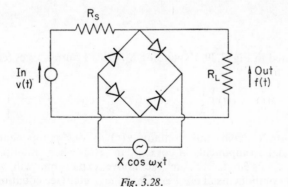

Fig. 3.28.

The output voltage is then given by $v_0 = R_L v_i/(R_L + R_S)$ and if the load R_L is much larger than the source resistance R_S we have $v_0 \approx v_i$. On the other half-cycle the diodes conduct with a forward resistance much smaller than R_S giving $v_0 \approx 0$. The output waveform $f(t)$ is therefore approximately related to the input waveform $v(t)$ as

$$f(t) = v(t) S(t) \qquad 3.41$$

where $S(t)$ is a 'switching function' which steps between 0 and 1 and back again at the frequency $\omega_X/2\pi$. The product of $v(t)$ and $S(t)$ is

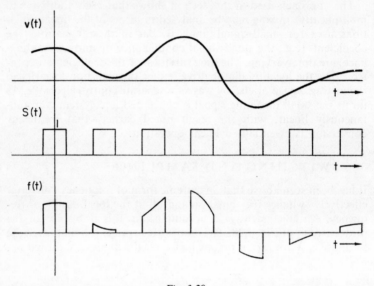

Fig. 3.29.

illustrated in Fig. 3.29. Expanding $S(t)$ as a Fourier series (equation 2.7) we have

$$f(t) = v(t)\left[\frac{1}{2} + \frac{2}{\pi}\cos\omega_X t - \frac{2}{3\pi}\cos 3\omega_X t + \ldots\right]$$

The output signal thus contains $v(t)$ as one component. Each sinusoidal component $A\cos\omega_A t$ in $v(t)$ when multiplied by $\cos\omega_X t$, $\cos 3\omega_X t$, etc., in the above expansion, will generate output components at $\omega_X \pm \omega_A$, $3\omega_X \pm \omega_A$, etc. (see equation 3.35).

104

The output spectrum is sketched in Fig. 3.30, and it is interesting to compare this with Fig. 3.22(b). This time there are no components on either side of the *even* harmonics of ω_X; the frequency-converted groups only appear about the odd harmonics ω_X, $3\omega_X$, etc. More significant is the fact that *no* components at the frequency ω_X and its harmonics appear in the output at all. Suppression of the local

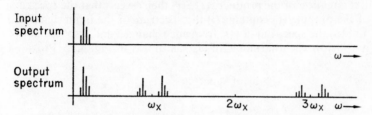

Fig. 3.30. *The input and output spectra for the circuit of Fig. 3.28.*

driving signal is typical of *balanced modulator* circuits of which Fig. 3.28 is one example, and is due to the balanced nature of the bridge arrangement. A somewhat more elaborate version is the circuit of Fig. 4.17, in which the signal is inverted on alternate half-cycles. The switching function $S(t)$ of equation 3.41 now steps between $+1$ and -1, so that there is no constant term in the expansion. This means that $v(t)$ is also suppressed from the output, leading to the term 'double-balanced ring modulator' for this arrangement.

The switching function need not be a square wave. The local oscillator in Fig. 3.28 may be replaced by a pulse generator so that $S(t)$ is now a train of narrow rectangular pulses as in Fig. 3.31.

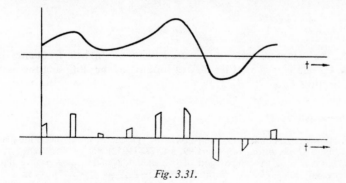

Fig. 3.31.

Signals and Information

The output is a sampled version of $v(t)$. If the sampling pulses are narrow, equation 2.9 may be taken as the spectrum of the switching function

$$S(t) = \frac{2\tau}{T}\left(\frac{1}{2} + \cos \omega_X t + \cos 2\omega_X t + \ \ldots\right)$$

Examination of the product $v(t)S(t)$ then reveals that the spectrum of the output $f(t)$ contains (i) the spectrum of the input signal $v(t)$ and (ii) the spectrum of $v(t)$ frequency changed and disposed above and below each harmonic of ω_X, as shown in Fig. 3.32. Provided

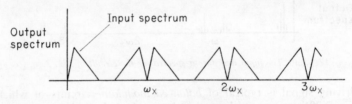

Fig. 3.32.

no overlapping occurs, the original signal $v(t)$ or some frequency-changed version of it may be extracted from $f(t)$ by filtering. Overlapping will occur, however, making it impossible to regain $v(t)$ unless

$$k\omega_X - W \geqslant (k-1)\omega_X + W$$

where k is an integer and W is the upper angular frequency limit of $v(t)$. This condition is seen to reduce to

$$\omega_X \geqslant 2W$$

or

$$f_X \geqslant 2F$$

i.e., the sampling rate must exceed twice the highest frequency in the signal. This is another way of looking at the sampling theorem of Section 2.15.

References

1. FLOYD, C. F., CORKE, R. L., and LEWIS, H., 'Design of linear phase low-pass filters', *Proc. Instn elect. Engrs* **99**, 3A, 777 (1952)
2. HOLT, A. G. J., 'A comparision of five methods of low-pass passive filter design', *Radio electron. Engng* **27**, 167 (1964)

106

3. STEINBERG, J. C., 'Effects of phase distortion on telephone quality', *Bell Syst. tech. J.* **9**, 550 (1930)
4. GUILLEMIN, E. A., *Communication Networks*, vol. 2 (Wiley, New York, 1935)
5. HALSEY, R. J. and WILLIAMS, H., 'The Birmingham–Manchester–Holme Moss television cable system', *Proc. Instn elect. Engrs*, **99**, 3A, 398 (1952)
6. ALLNATT, J. W., 'The delay equalization of the London–Birmingham television cable system', *Proc. Instn elect. Engrs* **99**, 3A, 338 (1952)
7. MACDIARMID, I. F., 'A testing pulse for television links', *Proc. Instn elect. Engrs* **99**, 3A, 436 (1952)
8. LEWIS, N. W., 'Waveform responses of television links', *Proc. Instn elect. Engrs* **101**, 258 (1954)
9. STRUTT, M. J. O., 'On conversion detectors', *Proc. Inst. Radio Engrs* **22**, 981 (1934)
10. 'Mixing Valves', *Wireless Engr* **12**, 59 (1935)
11. JAMES, E. G. and HOULDIN, J. E., 'Diode frequency changers', *Wireless Engr* **20**, 15 (1943)
12. KLEINMAN, H. M., 'Application of dual gate MOSFET's in practical radio receivers', *Inst. elect. electron. Engrs Trans.* **BTR 13** No. 2 (1967)

PROBLEMS

1. A rectangular pulse of amplitude 10 V and duration 1 μs is applied to a low-pass filter with a cut-off frequency of 10 kHz. What is the approximate amplitude and half-height duration of the output pulse?

2. The input pulse of Problem 1 is passed through a system whose bandwidth extends to 10 MHz. What will be, approximately, (a) the rise time of the output pulse, (b) the maximum rate of rise of the output voltage? The moment of arrival of the pulse is recorded as the instant at which the output signal first reaches 5 V. Unknown to the observer the signal has been lifted by 0·5 V, due to noise and drift, so that the rise is from 0·5 to 10·5 V. What error will be made in the timing? Show that this error is inversely proportional to the transmission bandwidth.

3. A certain network has an impulse response given by $h(t) = ct (0 < t \leqslant T)$, $h(t) = c(2T - t)(T < t \leqslant 2T)$, $h(t) = 0$ $(t > 2T)$. Find, by convolution, the response to a rectangular pulse of duration $T_1 (T_1 > 2T)$.

4. Use equation 3.26 to find the transfer function of the 'time-averaging' filter of Section 3.11.

5. Consider, by convolution, the response of a 'time-averaging' filter to sinusoidal input signals. Hence justify the statement in

107

Section 3.12 that the response to a sine wave at frequency f will be small for $f > 1/2T$. At what frequencies will the response be zero? Compare with the result of Problem 4.

6. Is the arrangement here a linear system? Show, by considering its impulse response, that it is a time-averaging filter.

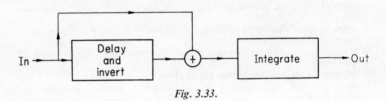

Fig. 3.33.

7. A certain non-linear network gives an output, v_o, which is related to the input, v_i, as

$$v_o = 10v_i + 0 \cdot 2v_i^2$$

Find the components of the output signal if the input is the sum of two sinusoids with amplitudes 0·2 and 0·5 V and frequencies 1 kHz and 200 Hz, respectively.

8. Verify that the system described by equation 3.36 has the conversion efficiency given by equation 3.37, making use of the trigonometric identities

$$\cos^2 \theta = \tfrac{1}{2} \cos 2\theta + \tfrac{1}{2}$$

$$\cos^3 \theta = \tfrac{1}{4} \cos 3\theta + \tfrac{3}{4} \cos \theta$$

$$\cos^4 \theta = \tfrac{1}{8} \cos 4\theta + \tfrac{1}{2} \cos 2\theta + \tfrac{3}{8}$$

$$\cos^5 \theta = \tfrac{1}{16} \cos 5\theta + \tfrac{5}{16} \cos 3\theta + \tfrac{5}{8} \cos \theta$$

$$\cos^6 \theta = \tfrac{1}{32} \cos 6\theta + \tfrac{3}{16} \cos 4\theta + \tfrac{15}{32} \cos 2\theta + \tfrac{5}{16}$$

$$\cos^7 \theta = \tfrac{1}{64} \cos 7\theta + \tfrac{7}{64} \cos 5\theta + \tfrac{21}{64} \cos 3\theta + \tfrac{35}{64} \cos \theta$$

Chapter 4

Modulation Theory

4.1 INTRODUCTION

It can be shown that in order to radiate efficiently an electro-magnetic wave of wavelength λ, the aerial employed must have a length which is comparable to or greater than $\lambda/4$. For a frequency of 1 kHz this dimension is approximately 70 km, so that it would be extremely inconvenient to broadcast speech or music signals by radiating them directly. Even if it were attempted the radiation efficiency would vary considerably within the signal bandwidth, being high at, say, 1 kHz but very poor at 100 Hz. On the other hand, a sine wave at 1 MHz may be efficiently radiated from a conveniently short aerial and the impedance and general properties of the aerial would change very little over a band of frequencies within 10 kHz or so of 1 MHz, as this now represents only a 1% change. One method of using radio waves to transmit signals, is, therefore, to frequency-convert the signal to a much higher frequency at which it may be radiated easily. The frequency at which the signal is relocated must be large compared with the signal bandwidth, so that the radiating characteristics of the aerial are sensibly the same for all components. This accounts for the fact that television broadcast frequencies lie at tens or hundreds of MHz whereas speech or music transmissions may be located as low as 100 kHz.

Modulation may be defined as the processing of a 'raw' signal to produce a new signal which is better adapted to the characteristics

109

of a communication channel. The example above shows that frequency conversion is one form of modulation and will be discussed later as 'single sideband modulation'. At the receiving end the reverse process of *demodulation* (or *detection*) is employed to recover the signal in its original form.

The advantages of frequency conversion as a modulation technique are not restricted to radio communications. As a second example we may consider the transmission of a television video signal over a long cable. Reference to Fig. 3.17 shows that direct transmission would result in severe phase distortion due to the steep slope of the group delay characteristic at low frequencies. The group delay of a typical coaxial cable for this application becomes fairly constant above about 500 kHz so that in practice frequency translation can be used to lift the lowest component of the video signal from almost zero to approximately 500 kHz. Because the attenuation of a cable rises with increasing frequency this means that the total available bandwidth is somewhat reduced, but the remaining phase distortion can be more easily removed by compensating networks, giving a better overall picture quality than would be possible with direct transmission. Some details of a practical system employing this technique are given in reference 1.

Quite apart from the technical advantages to be gained from modulation, there is an important economic advantage, namely the ability to use the same communication link to carry several messages simultaneously, a facility known as *multiplexing*. Several telephone conversations may be transmitted over a single cable, or reflected from the same satellite, by locating each in a different part of the frequency spectrum. Band-pass filters are then used at the receiving end to separate the channels prior to demodulation. The division of the frequency axis into separate channels in this way is termed *frequency division multiplexing*. Pulse modulation methods also permit multiplexing, by interleaving on the time axis the pulses corresponding to different messages, an arrangement called *time division multiplexing*. In all cases multiplexing requires an increased bandwidth, roughly proportional to the number of channels accommodated. Provided this additional bandwidth is available, multiplexing offers an economically useful flexibility whenever the ratio of the cost of the link to the cost of the additional terminal equipment is high.

The emphasis in this chapter will be on a description of the signals produced by the more common modulation methods and an examination of their bandwidth requirements. An important consideration in the choice of a particular modulation method will also be the signal-to-noise ratio achieved. A comparison of modulation methods from this point of view will be delayed until Chapter 7.

4.2 AMPLITUDE MODULATION

Amplitude modulation has been widely used in radio communications, and, because of the relative simplicity of the demodulation process in the receiver, it is still commonly employed in broadcasting. The process is illustrated in Fig. 4.1 in which the amplitude

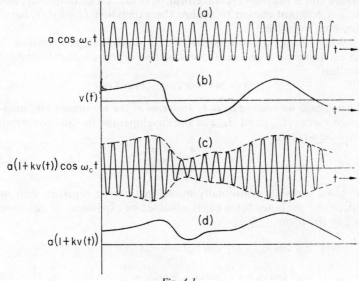

Fig. 4.1.

of a high-frequency sinusoidal *carrier* shown at (a) is made to vary in sympathy with the modulating signal (b). The resulting signal, shown at (c), is an amplitude-modulated or a.m. signal which carries the message waveform (b) as its envelope. On reception of the waveform (c), the primary function of the receiver is to provide an output proportional to the amplitude of the received carrier. The result, shown at (d), is identical to the modulating signal (b) apart from the added d.c. level which is usually of no consequence and may be removed.

111

Signals and Information

Since demodulation proceeds by recovery of the envelope, two conditions need to be fulfilled. First, the modulating signal must not change significantly between the individual maxima or minima of the carrier. This requires a carrier frequency much higher than the highest frequency in (b). Secondly, the negative-going excursions of the modulating signal must not be so great as to drive the envelope 'below zero'. A negative carrier amplitude has no meaning. The equation of the waveform (c) may be written

$$f(t) = a(1+kv(t)) \cos \omega_c t \qquad 4.1$$

where $v(t)$ is the message waveform, ω_c is the carrier frequency, and k is a constant chosen to ensure the expression $(1+kv(t))$ never becomes negative.

When the modulating signal $v(t)$ is sinusoidal, the constant k in equation 4.1 is termed the *modulation depth*, usually designated m, thus

$$f(t) = a(1 + m \cos \omega_m t) \cos \omega_c t \qquad 4.2$$

In this case we require $m \leqslant 1$. In terms of the maximum and minimum values A_{max} and A_{min} of the envelope, the modulation depth may be written

$$m = \frac{A_{max} - A_{min}}{A_{max} + A_{min}} \qquad 4.3$$

Fig. 4.2 shows a sinusoidally modulated carrier together with its spectrum, the latter being easily obtained by expansion of equation 4.2 in the form

$$f(t) = a \cos \omega_c t + am \cos \omega_c t \cos \omega_m t$$
$$= a\left[\cos \omega_c t + \frac{m}{2} \cos (\omega_c + \omega_m)t + \frac{m}{2} \cos (\omega_c - \omega_m)t\right] \qquad 4.4$$

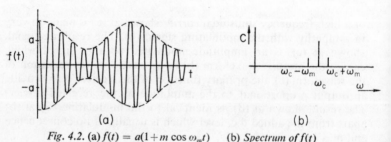

Fig. 4.2. (a) $f(t) = a(1 + m \cos \omega_m t)$ (b) *Spectrum of $f(t)$*

There are three components. One of amplitude a at carrier frequency and two so-called *sideband terms* of amplitude $am/2$ at $\omega_c \pm \omega_m$. These three components are orthogonal (over a sufficiently long period) so that the mean power in the modulated signal is

$$P_m = \frac{a^2}{2} + \frac{a^2 m^2}{8} + \frac{a^2 m^2}{8}$$

$$= P_0\left(1 + \frac{m^2}{2}\right) \qquad 4.5$$

where P_0 represents the power in the unmodulated carrier. Notice that even for a fully modulated wave ($m = 1$), only one-third of the total power is carried by the sidebands—i.e. in the message-bearing part of the spectrum—while two-thirds of the power is spent on the carrier. Note also that the bandwidth occupied by $f(t)$ is twice the frequency ω_m. Means of overcoming this waste of power and bandwidth will be discussed below (Section 4.6).

When $v(t)$ is a periodic waveform then it is easily seen that each of its Fourier components will produce sidebands above and below the carrier and separated from it by the frequency of that component. If $v(t)$ is a transient or pulse signal with Fourier transform $V(\omega)$ then expansion of equation 4.1, i.e.

$$f(t) = a \cos \omega_c t + akv(t) \cos \omega_c t$$

shows that $F(\omega)$ contains a pair of δ-functions of strength πa at $\pm \omega_c$ due to the carrier term $a \cos \omega_c t$ (Fig. 2.15 J), together with a 'sideband component' which is the spectrum of

$$f_s(t) = akv(t) \cos \omega_c t$$

namely

$$F_s(\omega) = ak \int_{-\infty}^{\infty} v(t) \cos \omega_c t \, e^{-j\omega t} \, dt$$

$$= ak \int_{-\infty}^{\infty} v(t) \frac{e^{j\omega_c t} + e^{-j\omega_c t}}{2} e^{-j\omega t} \, dt$$

$$= \frac{ak}{2} V(\omega - \omega_c) + \frac{ak}{2} V(\omega + \omega_c) \qquad 4.6$$

This result is sketched in Fig. 4.3, which shows the spectra of the modulating pulse and of the corresponding a.m. waveform.

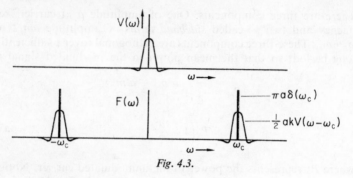

Fig. 4.3.

4.3 GENERATION AND DEMODULATION OF A.M. SIGNALS

Returning for a moment to Fig. 3.22, it may be seen that if the 'incoming signal' is replaced by a message waveform $v(t)$ and the local oscillator signal is regarded as a carrier which is being mixed with $v(t)$ in a non-linear device, then a filter with a pass-band from $\omega_X - \omega_2$ to $\omega_X + \omega_2$ would select an amplitude modulated version of $v(t)$ from the output. It is sufficient to have a quadratic term in the non-linear characteristic in order to produce this a.m. component and a modulator working on this principle is generally termed a *square law modulator*. Typically $v(t)$ might be a speech signal with a spectrum from 300 Hz to 3 or 4 kHz so that, if the carrier frequency exceeds 100 kHz or so, both the upper and lower sidebands would lie within the bandwidth of a simple resonant circuit tuned to the carrier frequency. Fig. 4.4 shows a simplified circuit diagram

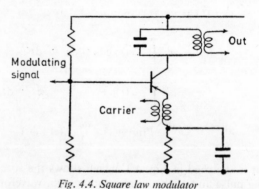

Fig. 4.4. Square law modulator

of a square law modulator, the bias condition being chosen to give non-linear operation.

The depth of modulation obtainable is not large because the carrier component in the output comes from the linear term, whereas the sidebands come from the quadratic term which is usually much smaller. If only linear and quadratic terms were present this could be overcome by increasing the amplitude of $v(t)$ relative to the input carrier. In practice this leads to distortion of the output envelope due to the presence of cubic and higher terms (the condition $X \gg A$ of Section 3.13 being violated).

The above method is unsuitable for high-power applications on account of its poor efficiency. The most efficient r.f. power ampifiers are run under class C conditions. By suitable choice of load and operating conditions it is possible to obtain a linear relationship between the h.t. potential and r.f. output voltage of a class C amplifier. It is then possible to use the valve as a modulated amplifier by arranging for the modulating signal to vary the h.t. supply voltage as shown in Fig. 4.5. The linearity means that the envelope faith-

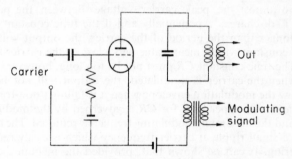

Fig. 4.5. Anode-modulated class C amplifier

fully follows the modulating waveform $v(t)$. Modulators such as this which do not rely on non-linear mixing to generate the sideband components, are classed as *linear modulators*. The circuit of Fig. 4.5 can work efficiently at high power and may form the output stage of a radio transmitter. This method is preferable to modulating at low power and amplifying a modulated carrier, since preservation of the envelope shape would then require linear power amplifiers which are relatively inefficient. It is perhaps worth noting, however, that the act of modulation increases the output power level (see equation 4.5). In Fig. 4.5 this additional power comes from the

115

modulating signal drive, which must therefore be supplied from a high-power linear amplifier. This objection may be overcome by modulating at the grid rather than at the anode, but the reader is referred elsewhere for futher details.

The process of demodulation or 'detection' may be satisfactorily accomplished using the diode detector circuit of Fig. 4.6. Some version of this circuit is used in almost all a.m. radio receivers. On the

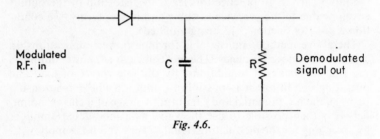

Fig. 4.6.

positive peaks of the carrier the diode conducts, charging the capacitor C to almost the peak carrier voltage. Between the positive peaks, C discharges exponentially and if the time constant CR is much longer than the period of the carrier, the output voltage is almost constant with a mean value very close to the carrier amplitude. The time constant CR must not be too long, however, otherwise, when the carrier is modulated, the output will not be able to follow the modulation envelope when it is falling ('non-tracking' distortion). An upper limit for CR is governed by the modulation depth and the maximum modulating frequency required. The output carries a small ripple at carrier frequency which is easily removed by filtering. It can be shown that, provided the incoming carrier amplitude is large compared with the diode 'turn-on' voltage, the output is a linear function of the carrier amplitude and this circuit provides an almost ideal linear demodulator.

4.4 GENERAL DESIGN FEATURES OF AN A.M. RECEIVER

A radio receiver must accomplish the following tasks: (i) select the portion of the spectrum occupied by the desired signal, (ii) amplify the weak signal received to a level suited to the demodulator, (iii) demodulate and if necessary further amplify the signal to whatever output power level is needed.

Operations (i) and (ii) are usually combined by using frequency-selective amplifiers whose bandwidth is equal to that of the modulated signal received. This bandwidth is typically 10 kHz or so for broadcast purposes and is, of course, independent of carrier frequency. It is difficult, however, to design a stable amplifier whose pass band may be tuned across the spectrum while maintaining constant bandwidth. A simpler solution is to convert all incoming signals to a fixed *intermediate frequency* (i.f.). Most of the amplification and filtering is then performed at this frequency. A block diagram of this arrangement is given in Fig. 4.7. Designating the local oscil-

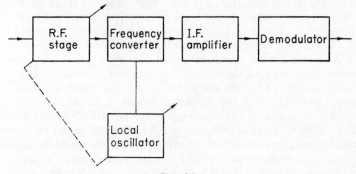

Fig. 4.7

lator frequency (or heterodyne frequency) as ω_h, an incoming carrier at ω_c will cause the frequency converter output to contain a component at the intermediate frequency ω_i, whenever the sum or difference of ω_c and $n\omega_h$ ($n = 1, 2, 3\ldots$) lies at ω_i. Many carriers may satisfy this condition simultaneously so that the frequency converter must be preceded by a tuned circuit, or tuned radio-frequency amplifier, which passes only the one desired. The input circuit and local oscillator are then varied together, in order to tune from one incoming carrier to another. The i.f. stages are invariably tuned to the difference frequency ($\omega_i = \omega_h - \omega_c$ or $\omega_i = \omega_c - \omega_h$) rather than to the sum, for then the i.f. may be chosen at a conveniently low value for achieving a high degree of stable amplification, even when ω_c is very high. It is also customary to locate the heterodyne frequency above the input circuit resonant frequency. In the m.f. broadcast band (500 to 1500 kHz) this is essential, for if the local oscillator were set *below* the carrier frequency by the usual i.f.

117

of 465 kHz a tuning range of from 35 to 1035 kHz would be required. This represents a ratio of 29·6 : 1 in frequency or 880 : 1 in tuning capacitance, which is impossible to achieve in practice owing to the stray capacitance which limits the minimum of the range. With $\omega_h > \omega_c$, on the other hand, the local oscillator must tune from 965 to 1965 kHz, representing a much more comfortable frequency ratio of about 2 : 1.

The intermediate frequency is chosen to be low enough to facilitate high-gain i.f. amplification. A low i.f. also eases the design of filter circuits having a sharp cut-off on either side of the pass band, this being necessary in order to reject signals occupying adjacent channels in the spectrum. On the other hand, ω_i must be large enough to avoid what is called 'image reception' or 'second-channel interference'. Image reception occurs when a carrier is present at $\omega_h + \omega_i$, as well as at $\omega_h - \omega_i$ (to which the receiver is tuned). It is, as explained above, the function of the input tuned circuits to reject this unwanted signal, but the frequency difference $2\omega_i$ between the desired carrier and its image may be insufficient if ω_i is small, particularly if the unwanted carrier is strong. An i. f. of 465 kHz is a satisfactory compromise for m.f. (0·3–3 MHz) reception, while 10 MHz is common in v.h.f. (30–300 MHz) receivers. Specialised communications receivers, which are required to operate on very weak signals, usually employ two stages of frequency conversion. The first has a high i.f. to give good image rejection and the second uses a low i.f. (about 100 kHz), commonly with crystal filters in the second i.f. amplifier, providing a sharply defined pass band to avoid interference from adjacent channels. One or more stages of r.f. amplification before the first mixer are usually incorporated, these being designed to give the best possible signal-to-noise ratio and freedom from cross-modulation (see Section 4.6 below).

4.5 TRANSMISSION OF A.M. SIGNALS THROUGH BAND-PASS FILTERS

The use of band-pass filters or amplifiers to separate the adjacent channels of radio- or other frequency multiplexed transmissions makes it necessary to consider the distortion of a.m. signals arising from this type of filtering. In the previous chapter special attention was paid to the transmission of waveforms through low-pass filters. It will now be shown how the results of that analysis may be extended to cover the closely analogous case of the transmission of an amplitude-modulated waveform through a band-pass filter, centred on the carrier frequency.

The incoming signal has the form of equation 4.1, with the spectrum sketched in Fig. 4.3. Excluding the δ-functions, which describe the carrier component, the spectrum (sideband spectrum) is given by equation 4.6

$$F_s(\omega) = \frac{ak}{2}[V(\omega-\omega_c)+V(\omega+\omega_c)]$$

where $V(\omega)$ is the transform of the modulating signal $v(t)$. In what follows it is assumed that $v(t)$ is band-limited to a relatively small fraction of the carrier frequency so that the two parts of $F_s(\omega)$ in equation 4.6 do not overlap. That is, the spectrum (as in Fig. 4.3) comprises two distinct peaks, one centred on $\omega = \omega_c$ and the other on $\omega = -\omega_c$.

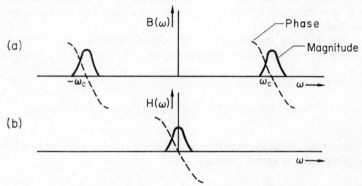

Fig. 4.8. (a) *Band-pass filter transfer function;* (b) *Equivalent low-pass filter*

The band-pass filter to which the signal is applied has a transfer function $B(\omega)$ which also possesses two distinct peaks at $\pm\omega_c$ as in Fig. 4.8(a). We assume that the peaks of $B(\omega)$ are symmetrical about the carrier frequency such that

$$B(\omega_c-\delta) = B^*(\omega_c+\delta) \qquad 4.7$$

Using the general relationship $B(\omega) = B^*(-\omega)$ inherent in the definition of the transfer function (equation 3.17), we may change the sign of the argument and take complex conjugates on both sides of equation 4.7 so that

$$B(-\omega_c+\delta) = B^*(-\omega_c-\delta) \qquad 4.8$$

The kind of symmetry expressed by equations 4.7 and 4.8 is exhibited by high-Q tuned circuits and coupled circuits. The assumed symmetry means that the two parts of the transfer function, at $\pm\omega_c$, are identical and not just mirror-images. Either peak may be described in terms of the function $H(\omega)$ sketched in Fig. 4.8(b) and we write

$$B(\omega) = H(\omega-\omega_c)+H(\omega+\omega_c) \qquad 4.9$$

It is assumed that $H(\omega) \equiv 0$ for $|\omega| > \omega_c$, a condition which ensures that the peaks in $B(\omega)$ are clearly separated. This means that when $\omega > 0$ the second term on the right of equation 4.9 vanishes, and it follows from equation 4.7 that

$$H(-\omega) = H^*(\omega) \qquad 4.10$$

The function $H(\omega)$ may thus be regarded as the transfer function of a hypothetical low-pass filter, which we shall call the *equivalent low-pass filter*.

The Fourier transform of the output signal may now be expressed, from equations 4.6 and 4.9, as

$$G_s(\omega) = B(\omega)\,F_s(\omega)$$

$$= \frac{ak}{2}\,[H(\omega-\omega_c)+H(\omega+\omega_c)]\,[V(\omega-\omega_c)+V(\omega+\omega_c)]$$

The narrowband assumptions about $H(\omega)$ and $V(\omega)$ ensure that the cross-products vanish, leaving

$$G_s(\omega) = \frac{ak}{2}\,[H(\omega-\omega_c)\,V(\omega-\omega_c)+H(\omega+\omega_c)\,V(\omega+\omega_c)]$$

$$= \frac{ak}{2}\,[G(\omega-\omega_c)+G(\omega+\omega_c)] \qquad 4.11$$

where
$$G(\omega) = H(\omega)\,V(\omega) \qquad 4.12$$

The carrier component, lying at the centre frequency of the filter, has been omitted from this spectrum, since it is assumed to be transmitted without attenuation. Clearly, equation 4.11, having the same form as equation 4.6, represents the sideband spectrum of the output signal, which is seen to be an amplitude-modulated waveform whose envelope has the spectrum $G(\omega)$ of equation 4.12. This last equation shows that the *output modulation g(t) would be the response of the equivalent low-pass filter to the input modulation* $v(t)$*.* This result is sometimes called the *low-pass/band-pass analogy.*

The above argument is recapitulated graphically in Fig. 4.9. The left-hand side of the figure, working from top to bottom, shows (a) the spectrum of the input a.m. wave $a(1+kv(t))\cos\omega_c t$, (b) a band-pass filter transfer function, (c) the spectrum of the output a.m. wave $a(1+kg(t))\cos\omega_c t$. The right-hand side of the figure

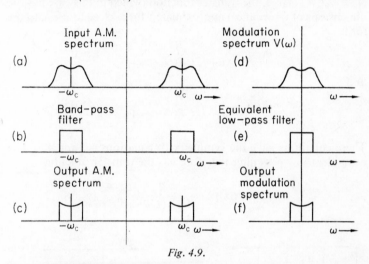

Fig. 4.9.

shows the corresponding low-pass operation. The equivalent low-pass filter is seen to attenuate (and change the phases of) the Fourier components of the modulating signal in just the same way as the band-pass filter operates on the *sidebands* of the a.m. signal. It would make no difference to the filtered signal whether the band-pass filter were inserted after modulation or the equivalent low-pass filter were inserted before modulation. The band-pass filtering of a.m. signals is therefore most easily considered in terms of equivalent low-pass filtering of the envelope. The following examples will illustrate this technique.

Example 1: Response of a tuned circuit to a suddenly applied sinusoid

The band-pass filter of Fig. 4.10 is a simple parallel *LCR* circuit. In order to find the transfer function we first consider a constant current source which supplies an input current $ie^{j\omega t}$ giving rise to an

9*

121

output voltage $v_o e^{j\omega t}$ which depends on frequency according to the familiar resonance curve

$$v_o = \frac{iR}{1+j\omega_0 CR(\omega/\omega_0 - \omega_0/\omega)}$$

where v_o is complex and ω_0 is the resonant frequency. When $Q = \omega_0 CR$ is large, the transfer function (which in this case has the dimensions of impedance) approximates, for ω close to $\pm\omega_0$, to the form

$$\frac{v_o}{i} = B(\omega) = \frac{R}{1+j2CR(\omega - \omega_0)} \qquad \omega > 0$$

and

$$B(\omega) = \frac{R}{1+j2CR(\omega + \omega_0)} \qquad \omega < 0$$

These equations have the symmetry of equations 4.7 and 4.8. The equivalent low-pass filter therefore has the transfer function

$$H(\omega) = \frac{R}{1+j2CR\omega}$$

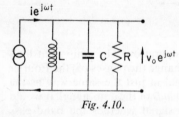

Fig. 4.10.

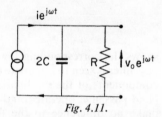

Fig. 4.11.

Now, it is easily shown that the circuit of Fig. 4.11 has precisely this transfer function and is thus the equivalent low-pass filter for this example.

Fig. 4.12(a) shows a step-function input current I applied to the equivalent low-pass filter. The response has the well-known form (b)

$$v = IR(1 - e^{-t/2CR})$$

It follows from the low-pass/band-pass analogy that if a current $I\cos\omega_0 t$ were to be applied at $t = 0$ to the circuit of Fig. 4.10, the

122

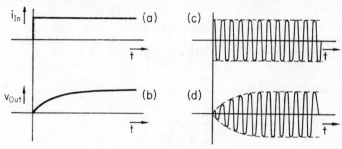

Fig. 4.12.

response would be as in Fig. 4.12(d), namely

$$v = IR(1 - e^{-t/2CR}) \cos \omega_0 t$$

This result agrees with the one obtained more laboriously by the conventional methods of circuit analysis. Note that the envelope rise time is given by $2CR = 2/\Delta\omega$, where $\Delta\omega = \omega_0/Q$ is the (angular-frequency) bandwidth between the -3 dB points. The method above may, of course, be used to discover the response of the tuned circuit to any modulated waveform (provided the carrier lies at its resonant frequency) by considering the response of the equivalent low-pass filter to the modulation envelope. Fig. 4.13 shows the

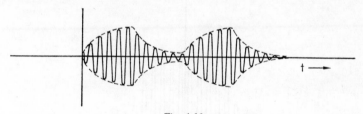

Fig. 4.13.

response to a pair of Morse dots of duration T. It is clear that these would not be well resolved if the rise time were too long, i.e. if the bandwidth were too small.

Example 2: Pulse response of an idealised band-pass filter

Fig. 4.14(a) shows the transfer function of an idealised band-pass filter, centred on the carrier frequency and with bandwidth β. The equivalent low-pass filter clearly has the transfer function of Fig.

123

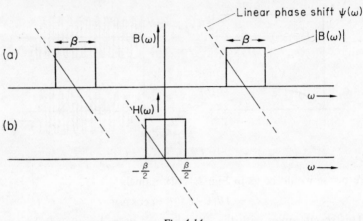

Fig. 4.14.

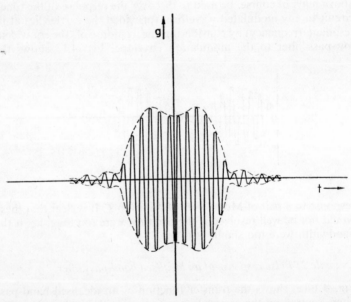

Fig. 4.15. Response of an idealised band-pass filter, bandwidth W Hz to a burst of oscillations of duration 4/W s, at mid-band frequency. Compare Fig. 3.13(b)

124

4.14(b) and is the idealised low-pass filter which was used as a model in Chapter 3. Its cut-off frequency lies at $\beta/2$ rad s^{-1}.

The response of such a filter to a burst of sinusoidal oscillations at mid-band frequency has the envelope of the response of the equivalent low-pass filter to a rectangular pulse of the same duration. This is illustrated in Fig. 4.15, which should be compared with Fig. 3.13(b). In Section 3.13 (see also Section 3.4) it was shown that for a pulse of duration τ to be transmitted through a low-pass filter without serious attenuation, the filter cut-off frequency must exceed $1/2\tau$. This result may be applied to the transmission of an a.m. pulse through a band-pass filter (e.g., the i.f. stages of a receiver) and we see that the filter bandwidth, β, must exceed $1/\tau$. To resolve two pulses separated by τ a bandwidth of approximately $1/\tau$ Hz is also needed. If the input envelope is a step function then equation 3.14 shows that the envelope of the response has a rise time $2\pi/\beta$. These are important results for deciding the bandwidth requirements for pulsed a.m. communication systems or for radar systems. The low-pass analysis of Chapter 3 is used, remembering that the equivalent low-pass bandwidth is $\beta/2$.

4.6 NON-LINEAR EFFECTS. SQUARE-LAW DETECTION. CROSS-MODULATION

Apart from amplitude and phase distortion due to linear filtering, an a.m. signal will also be subject to non-linear distortion due to slight curvature in the characteristics of amplifying devices. If ω_c, ω_u and ω_l are the carrier and upper and lower sideband frequencies, respectively, of a sinusoidally modulated signal, non-linear distortion will introduce intermodulation frequencies in the manner discussed in Section 3.12. Consider first the effect of a quadratic term. This will bring into existence components at $2\omega_c$, $2\omega_u$ and $2\omega_l$, together with the sums and differences $\omega_c \pm \omega_u$, $\omega_c + \omega_l$ and $\omega_u \pm \omega_l$. None of these, however, lies close to carrier frequency and they will be rejected by subsequent tuned circuits so that no distortion need result from the presence of a quadratic term. On the other hand, it is possible to *demodulate* the a.m. signal by means of a square-law device followed by a low-pass filter which passes only the terms at $\omega_u - \omega_c$, $\omega_c - \omega_l$, and $\omega_u - \omega_l$. The first two are identical and lie at the modulating frequency ω_m, the last one lies at twice the modulating frequency but will be relatively weak if the sideband amplitudes are small compared with the carrier (small modulation depth). The output is then a sine wave at the modulating frequency

125

with some second-harmonic distortion. Recovery of the modulating signal in this way is termed *square-law detection* (see also Section 3.12) and some early forms the demodulator—e.g., the 'anode bend detector'—operated on this principle.

The presence of a cubic term will give rise to components such as $\omega_c \pm \omega_u \pm \omega_l$. These include the frequencies $\omega_c \pm 2\omega_m$ which lie close to the carrier and will not be filtered out by the band-pass circuits. They represent second-harmonic distortion of the envelope.

A further unwanted phenomenon also arises due to the cubic term when an unmodulated carrier at ω'_e and a modulated carrier, with components ω_c, ω_u, and ω_l are both present at the input to a non-linear device. A subsequent band-pass filter centred on ω'_c will pass not only the carrier at ω'_c but also the intermodulation terms which lie at $\omega'_c \pm (\omega_u - \omega_c)$ and $\omega'_c \pm (\omega_c - \omega_l)$. Note that a cubic or higher term is needed to generate these components, which effectively form sidebands at $\omega'_c \pm \omega_m$. The carrier at ω'_c now appears to be modulated by the modulation belonging to the other carrier. This effect is termed *cross-modulation*. In practice it may occur when a radio receiver is tuned to a weak station while a strong station is transmitting on a neighbouring frequency. The first tuned circuits in the receiver may not be sufficiently selective to suppress the unwanted station, so that both are presented to the first valve or transistor. Cross-modulation at this point will give interference from the unwanted station which no amount of subsequent filtering can eliminate. The characteristic feature of this effect is that the interference ceases when the *wanted* station goes 'off the air'. Strict attention to linearity in the input stages of a communications receiver is necessary if weak signals are not to suffer cross-modulation from local broadcasting or other powerful signals.

4.7 SUPPRESSED-CARRIER A.M. SIGNALS

The carrier frequency component of an a.m. wave is wasteful of power since it carries no information; it is the sidebands which form the 'message' parts of the spectrum. Moreover, since these are mirror images of one another about the carrier frequency, only one set of sidebands need be transmitted. When the carrier is removed, the signal is called a *suppressed-carrier a.m. signal*. The signal is termed double-sideband suppressed-carrier a.m., or d.s.b., if both sets of sidebands are retained and single-sideband, or s.s.b., when only one set of sidebands, upper or lower, is retained. The latter variety uses only one-half the bandwidth of simple a.m. and has considerable appeal when an allotted transmission band must be utilised as fully as possible. Both d.s.b. and s.s.b. make efficient use of the transmitted power and, under certain conditions, it can be shown that s.s.b. gives a 9 dB better signal-to-noise ratio than conventional a.m. (see Section 7.7).

Mathematically we may write an s.s.b. signal, modulated by a sine wave at ω_m, as

$$f(t)_{\text{s.s.b.}} = a_1 \cos (\omega_c + \omega_m)t \quad \text{or} \quad a_1 \cos (\omega_c - \omega_m)t \quad 4.13$$

depending on whether the upper or lower sideband is used. The d.s.b. signal is

$$f(t)_{\text{d.s.b.}} = a_2[\cos (\omega_c + \omega_m)t + \cos (\omega_c - \omega_m)t]$$
$$= (2a_2 \cos \omega_m t) \cos \omega_c t \quad 4.14$$

The amplitude of the sidebands, a_1 or a_2, is in either case made proportional to the amplitude of the modulating signal. These signals are sketched and compared with the corresponding a.m. signal in Fig. 4.16.

In order to generate a suppressed carrier signal it is better to begin with a modulator in which the carrier is suppressed, rather than attempting to filter out the carrier from a normal a.m. wave.

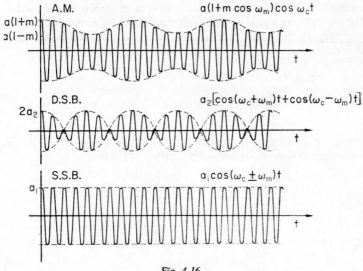

Fig. 4.16.

127

Signals and Information

A *balanced modulator* is used for this purpose and one variety of balanced modulator has already been discussed in Section 3.15. Fig. 4.17 shows another form of balanced switching modulator. The modulating signal $v(t)$ is applied to the primary of the first transformer while the driving oscillator, at carrier frequency, has a much larger amplitude and is connected across the centre taps

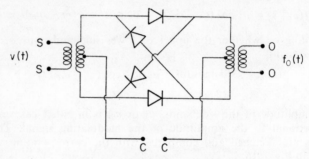

Fig. 4.17. A double-balanced modulator

as shown. The diodes are arranged so that, on alternate half-cycles of the carrier, the connections to the second transformer are reversed. The effect on the wave form is as shown in Fig. 4.18. The modulating signal is effectively multiplied by a switching function $S(t)$ which in this case steps from $+1$ to -1 and back again at carrier frequency. The output may then be written

$$f_o(t) = v(t)\, S(t)$$

$$= v(t) \times \frac{4}{\pi} \left[\cos \omega_c t - \frac{1}{3} \cos 3\omega_c t + \ \ldots \right]$$

where we have used the usual Fourier expansion for the square wave $S(t)$. When $v(t)$ is sinusoidal, $v(t) = a \cos \omega_m t$

$$f_o(t) = \frac{4a}{\pi} \cos \omega_m t \left[\cos \omega_c t - \frac{1}{3} \cos 3\omega_c t + \ \ldots \right]$$

$$= \frac{2a}{\pi} \cos (\omega_c + \omega_m)t + \frac{2a}{\pi} \cos (\omega_c - \omega_m)t$$

$$+ \frac{2a}{3\pi} \cos (3\omega_c + \omega_m)t + \ \ldots \qquad 4.15$$

128

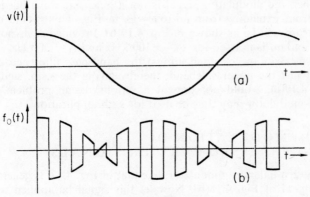

Fig. 4.18.

This is the signal sketched in Fig. 4.18(b) and is seen to contain side-bands about the carrier frequency and its odd harmonics. When $f_c(t)$ is passed through a band-pass filter centred on ω_c, the high-frequency terms are removed, leaving the d.s.b. signal of equation 4.14 (Fig. 4.16). Note that equation 4.15 does not contain terms at either the carrier or modulating frequencies.

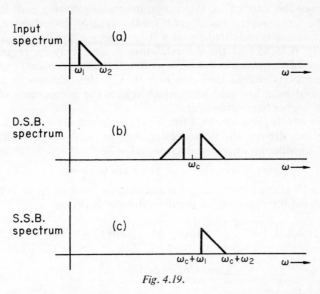

Fig. 4.19.

When the modulating signal is not a pure sine wave but has a spectrum extending from ω_1 to ω_2 as in Fig. 4.19(a), the d.s.b. spectrum will be as shown in Fig. 4.19(b). Provided ω_c is not too high and ω_1 is not too low ($f_c = 100$ kHz and $f_1 = 300$ Hz would be typical in speech applications) the band-pass filter needed to select just the upper sideband, thereby giving the s.s.b. signal of Fig. 4.19(c), would not present a serious design problem. The same modulator may thus be used for s.s.b. applications.

4.8 DEMODULATION OF SUPPRESSED-CARRIER SIGNALS

Suppose a balanced modulator, like that of Fig. 4.17, is generating the signal of Fig. 4.18(b). Now let this signal be applied to the terminals *s–s* of a second bridge which is identical to that of Fig. 4.17. A locally supplied sinusoid is applied across *c–c* as before. If this local 'carrier' has the same frequency and phase as the modulating carrier, then the switching action will be in perfect synchronism to re-invert alternate sections of the modulated waveform (Fig. 4.18(b)) and the output will be a replica of the modulating signal (Fig. 4.18(a)). A balanced modulator arrangement may thus also be used as a demodulator. The requirement of synchronism between the 'carriers' at the transmitting and receiving ends leads to the term *synchronous detector* for this type of demodulator. In practice, the transmitted signal will have been filtered to the form of Fig. 4.16(b) and the demodulation process may be regarded as one of frequency conversion in which the (suppressed) carrier frequency is shifted down to $\omega = 0$. The demodulator is then followed by a low-pass filter which rejects the components close to the higher harmonics of ω_c.

The effects produced by a frequency and phase error in the local oscillator driving the demodulator may be analysed as follows. First consider an incoming d.s.b. signal

$$f(t) = a \cos{(\omega_c + \omega_m)} t + a \cos{(\omega_c - \omega_m)} t$$

and a local oscillator running at frequency ω_c' with phase ϕ so that the switching function for the demodulator becomes

$$S(t) = \frac{4}{\pi} \left[\cos{(\omega_c' t + \phi)} - \frac{1}{3} \cos{(3\omega_c' t + 3\phi)} \right.$$
$$\left. + \frac{1}{5} \cos{(5\omega_c' t + 5\phi)} + \ldots \right]$$

The demodulator output is found by examining the product $f(t) S(t)$ and picking out the low-frequency components. This filtered output is easily shown to be

$$v_o(t) = \frac{2a}{\pi} \{\cos [(\omega_c' - \omega_c - \omega_m)t + \phi]$$

$$+ \cos [(\omega_c' - \omega_c + \omega_m)t + \phi]\} \qquad 4.16$$

$$= \frac{4a}{\pi} \cos (\varepsilon t + \phi) \cos \omega_m t \qquad 4.17$$

in which the frequency error has been written $\varepsilon = \omega_c' - \omega_c$. When ε is not zero the output is seen to exhibit 'beats' at the error frequency as shown in Fig. 4.20. When $\varepsilon = 0$ no beats occur and the phase error ϕ is seen to reduce the output by a factor $\cos \phi$.

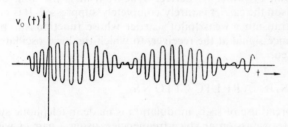

Fig. 4.20. The effect of frequency and phase errors in the local carrier in d.s.b. demodulation

A phase error of $\pi/2$ would give zero output, while a phase error π inverts the signal. It appears that a frequency error would give distortion in the form of unpleasant 'beats', while a small but constant phase error could be tolerated.

The same demodulator may, of course, be used for incoming s.s.b. signals, in which case there is only one sideband in the input signal and thus only one of the terms in equation 4.16 is retained. The output from the demodulator is now, assuming the input to be the upper sideband

$$v_o(t) = \frac{2a}{\pi} \cos [(\omega_m - \varepsilon)t - \phi] \qquad 4.18$$

The effect of a frequency error is, in this case, to decrease (or increase) the frequency of each component in the modulating signal by

131

the same amount. The effect is difficult to describe but is *not* the same as hearing a gramophone record played at the wrong speed. In the latter case every component frequency is changed by the same *factor* so that the harmonic relationships are unaltered and —except to listeners with 'perfect pitch'—a small error is not unpleasant. In the present situation, however, the effect is a linear frequency translation and, for speech communication, noticeable distortion results when the locally supplied carrier is in error by as little as ± 15 Hz. This represents a stringent requirement, especially as carrier frequencies may be tens of MHz.

If the frequency error is zero, the effect of a phase error is to change the phase of every Fourier component of the modulating signal by the same amount. This violates the condition for distortionless transmission and wave shapes will be distorted. For telephony applications, however, quite large phase errors can be tolerated. But, in principle, perfect synchronism is the aim, and for this reason the carrier is rarely completely suppressed. It is customary to transmit a weak 'pilot' carrier whose function is to provide a reference signal at the receiver, to which the local oscillator may be locked.

4.9 S.S.B. APPLICATIONS

Widespread use of s.s.b. modulation is made in telephone systems. Direct transmission at voice frequencies, using a pair of wires for each subscriber, can only be feasible at the local level between the subscriber and his exchange and between minor exchanges. Major zone exchanges are interconnected by coaxial cables which are usable at frequencies ranging up to several MHz or by other wideband links such as microwave links. Trunk telephone traffic is arranged by dividing the available spectrum into many channels, each 4 kHz wide.

A set of 12 channels is called a 'group'. Each of 12 speech signals is first passed through a filter which restricts the bandwidth to the range 300–3 400 Hz. Twelve carriers, with frequencies from 64 to 108 kHz spaced 4 kHz apart, are used to s.s.b.-modulate the individual signals. The lower sidebands are selected using band-pass filters and these are combined to yield the 'basic C.C.I.T.T.[†] 12-channel group' whose spectrum extends from 60 to 108 kHz. The choice of carrier frequencies was originally governed by the

† International Telephone and Telegraph Consultative Committee

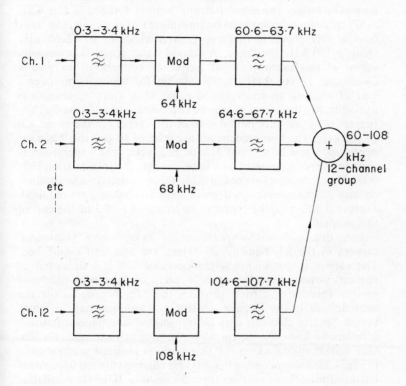

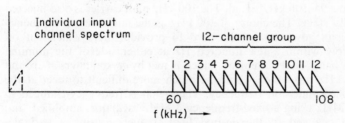

Fig. 4.21. Production and spectrum of a basic 12-channel group

133

available crystal filter techniques which are well adapted to this frequency range. The general arrangement is outlined in Fig. 4.21.

When a single group is to be transmitted over a cable, the basic 60–108 kHz group is frequency-converted down to 12–60 kHz using a 120 kHz oscillator. A second basic group is commonly added at this point (without frequency conversion) to give a 24-channel 12–108 kHz signal suitable for transmission over a pair of wires in an underground cable. If a wider bandwidth is available, five basic groups may be combined into a 60-channel 'supergroup' by the s.s.b. modulation of five carriers in the range 420 to 612 kHz. The lower sidebands are selected to give the 'basic supergroup' a spectrum from 312 to 552 kHz. The supergroup is then frequency-converted to a band suitable for transmission (60–300 kHz is common for coaxial cable transmission) or else combined with other supergroups if more than 60 channels are required. Further details may be found in references 2 and 3 at the end of this chapter.

Long-distance point-to-point radio links, using suppressed carriers in the h.f. band (3–30 MHz), are also well established. The carrier is not completely suppressed but is transmitted at reduced power, typically some 26 dB below the peak sideband power. This 'pilot carrier' provides the reference signal for the demodulator in the receiver, so that synchronism may be maintained. Several channels may be transmitted simultaneously using a technique known as *independent sideband* or i.s.b. transmission. One system employed on B.P.O. circuits is outlined schematically in Fig. 4.22. Two incoming 0·3–6·0 kHz channels are fed to separate balanced modulators driven from a common 100 kHz oscillator. The modulators are followed by filters which select the upper and lower sidebands respectively and their outputs are summed to yield a 94–106 kHz signal. The 100 kHz pilot carrier is also injected at this stage. The choice of 100 kHz as the initial carrier frequency is again governed by the need to provide filters which cut off sharply within a few hundred Hz at either side of the nominal pass band—a requirement which is met by crystal filters at around 100 kHz but which would be much more difficult to meet at the final transmission frequency. The signal is frequency-changed to 3·1 MHz using a fixed-frequency 3 MHz oscillator, amplified, and then changed to the desired final frequency using a variable-frequency oscillator before being supplied to the output power amplifier (which must be linear to avoid cross-modulation). Frequency conversion like this, in two stages, eases the problems

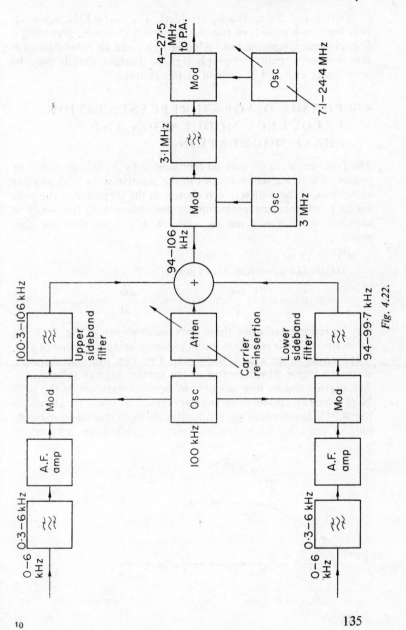

Fig. 4.22.

of filtering out the unwanted products. The two 6 kHz input signals may each consist of two 3 kHz speech channels, providing a four-channel telephone link. Alternatively, one or both sidebands may comprise many telegraph signals. Further details may be found in reference 4 at the end of this chapter.

4.10 PHASOR DIAGRAM REPRESENTATION. FREQUENCY MODULATION AND PHASE MODULATION

The function $a \cos \omega t$ may be represented by a rotating vector or *phasor* of length a which rotates in the positive sense with angular velocity ω. If the x-direction is chosen as the direction of the phasor at $t = 0$ then the projection of the phasor onto the x-axis at subsequent times is given by $a \cos \omega t$. It follows that the a.m. signal

a.m.:

$$f(t) = a(1 + m \cos \omega_m t) \cos \omega_c t$$

$$= a \cos \omega_c t + \frac{am}{2} \cos (\omega_c + \omega_m)t + \frac{am}{2} \cos (\omega_c - \omega_m)t$$

may be represented using three phasors, as shown in Fig. 4.23, the projection of their resultant onto the x-axis being the value of $f(t)$.

It is easier to see what is happening if we view this system from a reference frame which rotates at the carrier angular velocity, ω_c. The carrier phasor now appears to be stationary, while the sideband phasors rotate with angular velocities $\pm \omega_m$ as shown in Fig. 4.24. The length of the resultant gives the instantaneous amplitude of the a.m. wave, which is seen to vary from $a(1 + m)$ to

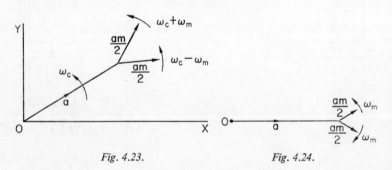

Fig. 4.23.　　　　　　　Fig. 4.24.

$a(1-m)$ as expected. The *direction* of the resultant, however, is constant, showing that the carrier phase and frequency are not varied by this form of modulation.

Fig. 4.25, on the other hand, shows a phasor whose *length* is constant but whose direction varies sinusoidally, as its end point shifts along the arc ABC and back again at the modulating fre-

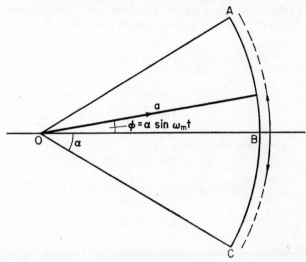

Fig. 4.25.

quency. Writing the phase angle referred to an unmodulated carrier OB as $\phi = \alpha \sin \omega_m t$, then this *phase-modulated* signal may be written

p.m.:
$$f(t) = a \cos (\omega_c t + \alpha \sin \omega_m t) \qquad 4.19$$

in which the maximum phase shift, α, is a parameter of the design of the system and has no simple upper limit (unlike the modulation depth in a.m. signals). The maximum excursion, α, measured in radians, is termed the *modulation index* of the p.m. signal and is arranged to be proportional to the amplitude of the modulating signal.

Because the phase shift ϕ is varying, the rocking phasor of Fig. 4.25 has, at any instant, a certain angular velocity with respect to OB which represents the instantaneous angular frequency deviation of the signal from the unmodulated frequency ω_c. This means the signal of equation 4.19 also exhibits *frequency modulation*. Frequency modulation and phase modulation always occur together. In phase modulation the instantaneous phase deviation φ is made to vary in sympathy with the modulating signal. Frequency modulation is somewhat different, however, for here it is the instantaneous frequency deviation, $d\varphi/dt$, which is made proportional to the modulating signal. Thus, for a sinusoidal modulating signal which is producing a *peak frequency deviation D*

$$\frac{d\varphi}{dt} = D \cos \omega_m t$$

thus

$$\varphi = \frac{D}{\omega_m} \sin \omega_m t$$

and hence we have for the f.m. signal

f.m.:

$$f(t) = a \cos \left(\omega_c t + \frac{D}{\omega_m} \sin \omega_m t \right) \qquad 4.20$$

$$= a \cos \left(\omega_c t + \beta \sin \omega_m t \right) \qquad 4.21$$

where β is the modulation index which, as in equation 4.19, represents the maximum phase deviation produced by the modulation. The distinction between p.m. and f.m. lies only in the fact that in the former the modulation index is independent of the modulating frequency, while in the latter it is inversely proportional to the modulating frequency. This makes a difference to the variation of bandwidth with modulating frequency, as will be seen below.

Both p.m. and f.m. signals may be represented by the phasor diagram of Fig. 4.25. That this 'rocking' phasor may be formed as the resultant of several component phasors having fixed frequencies (as in the a.m. case, Fig. 4.24) is less obvious, however. We shall see in the next section that, in general, so many components are required that their phasor representation is cumbersome. One simple case, however, is worth examination. This occurs when the modulation index α or β is a small angle. Consider the three component phasors of Fig. 4.26. A 'carrier' phasor $a \cos (\omega_c t + \pi/2)$ has been represented by the vertical line OB of length a. An upper

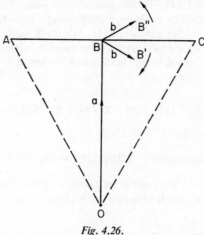

Fig. 4.26.

sideband $b \cos (\omega_c + \omega_m)t$ and a lower sideband $b \cos (\omega_c - \omega_m)t$ are represented by **BB'** and **BB''**, respectively. As the sideband phasors rotate in opposite directions (recalling that **OB** is stationary in this frame of reference) it is seen that the end point of the resultant phasor tracks backward and forward along the line **ABC**. The peak phase deviation is given by

$$\alpha = \tan^{-1} \frac{2b}{a}$$

and, when this is small, the amplitude changes represented by the difference in length of **OA** or **OC** and **OB** become negligible. The line **ABC** approximates to the arc of a circle, as in Fig. 4.25, and the modulation index is given approximately by $\tan^{-1}(2b/a) \approx 2b/a$. Phase or frequency modulation is termed *narrowband* when the modulation index is much less than unity. We now see that *narrowband* f.m. (or p.m.) contains three components, one at carrier frequency ω_c, together with two much weaker sidebands at $\omega_c \pm \omega_m$. The bandwidth needed for narrowband f.m. is $2\omega_m$, just as for simple a.m. Note that this is independent of the peak frequency deviation, so long as $\beta \ll 1$. This result is intriguing for it means that, to take an example, if the peak frequency deviation is 1 Hz and the modulating frequency is 1 kHz so that β is small ($\beta = 0.001$ in this case), the f.m. signal contains two

139

sidebands at ± 1 kHz from carrier frequency. Even though the instantaneous carrier frequency only sweeps through ± 1 Hz, the bandwidth of the signal is 2 kHz. We shall see below that more sidebands are needed as β grows larger and it is most important to realise that the signal bandwidth is not the same thing as the peak frequency deviation of an f.m. signal.

4.11 SPECTRA OF F.M. AND P.M. SIGNALS

Expanding equation 4.21 yields

$$f(t) = a[\cos \omega_c t \cos (\beta \sin \omega_m t) - \sin \omega_c t \sin (\beta \sin \omega_m t)] \quad 4.22$$

Further expansion is possible by using a result from the theory of Bessel functions which expresses $\cos (\beta \sin \theta)$ and $\sin (\beta \sin \theta)$ in the form

$$\cos (\beta \sin \theta) = J_0(\beta) + 2[J_2(\beta) \cos 2\theta + J_4(\beta) \sin 4\theta$$
$$+ J_6(\beta) \sin 6\theta + \ldots] \quad 4.23$$
$$\sin (\beta \sin \theta) = 2[J_1(\beta) \sin \theta + J_3(\beta) \sin 3\theta + J_5(\beta) \sin 5\theta + \ldots] \quad 4.24$$

where $J_n(\beta)$ is a Bessel function of order n and argument β.
    Substituting 4.23 and 4.24 into 4.22 yields

$$f(t) = a[J_0(\beta) \cos \omega_c t + J_1(\beta) \cos (\omega_c + \omega_m)t - J_1(\beta) \cos (\omega_c - \omega_m)t$$
$$+ J_2(\beta) \cos (\omega_c + 2\omega_m)t + J_2(\beta) \cos (\omega_c - 2\omega_m)t$$
$$+ J_3(\beta) \cos (\omega_c + 3\omega_m)t - J_3(\beta) \cos (\omega_c - 3\omega_m t)$$
$$+ \ldots] \quad 4.25$$

The spectrum of an f.m. (or p.m.) signal which is sinusoidally modulated at frequency ω_m to a modulation index β is thus seen to comprise a carrier component whose amplitude is $aJ_0(\beta)$ together with sidebands at intervals ω_m on either side of the carrier frequency. The nth sidebands, at $\omega_c \pm n\omega_m$, have amplitudes $aJ_n(\beta)$. Fig. 4.27 shows the form of $J_0(\beta)$, $J_1(\beta)$ and $J_2(\beta)$. It is seen that, as β increased from zero, the component at carrier frequency falls in amplitude, passes through zero for $\beta = 2 \cdot 405$, and subsequently rises and falls in an oscillatory manner. The amplitudes of the first, second, and higher-order sidebands are zero for $\beta = 0$. As β is increased the sidebands appear in sequence and the amplitude of each subsequently exhibits slowly decaying oscillations.

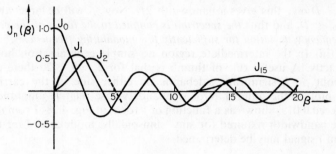

Fig. 4.27.

Although there is strictly an infinite number of sidebands, the amplitudes of only a finite number are significant. Referring to the table in Appendix 2 and taking as an example the case $\beta = 2$ it is seen that the relative amplitude of the carrier is 0·224, that of the first sidebands is 0·577, while the amplitudes of all subsequent sidebands steadily diminish, those beyond the sixth being all less than 0·001. For other values of β it is also seen that there is a limit for n beyond which the sideband amplitudes become negligible. This means that in practice an f.m. signal may be regarded as having a finite bandwidth.

The mean power of the signal of equation 4.21 is clearly $a^2/2$. In terms of the components of equation 4.25 we have

$$P = \frac{a^2}{2} = \frac{a^2}{2} \, [J_0^2(\beta) + 2J_1^2(\beta) + 2J_2^2(\beta) + \ldots]$$

The term in square brackets must therefore have the value unity. As a practical guide it is common to ignore sideband components for which $J_n(\beta) < 0.01$. This is seen to be equivalent to neglecting components which contribute 0·02% or less to the overall power. With this criterion in mind, a guide to the *number of significant sidebands* may be given as follows.

(i) $\beta \lesssim 0.3$. Only the first sideband on either side of the carrier is significant. This is the *narrowband* situation discussed at the end of the previous section.

(ii) $\beta \gtrsim 30$. At the other extreme, when β is sufficiently large, it can be shown that the number of significant sidebands on either side of the carrier is given roughly by $N = \beta$. The bandwidth is then $2\beta\omega_m$, where ω_m is the modulating frequency. Recalling that

141

$\beta = D/\omega_m$, this gives a bandwidth $2D$. Now, β will be large when $\omega_m \ll D$, and thus *the spectrum is confined to the limits of the peak frequency deviation for sufficiently low modulating frequencies.*

(iii) In the intermediate region no simple generalisation holds exactly. A useful rule of thumb is that for $2 < \beta < 20$ there are about $\beta + 3$ significant sidebands on either side of the carrier. A plot of the number N of sidebands whose relative amplitudes exceed 0.01 is shown as a function of β (or α) in Fig. 4.28. From this the bandwidth required for any sinusoidally modulated f.m. (or p.m.) signal may be determined.

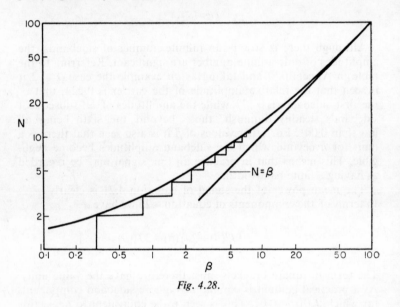

Fig. 4.28.

The sideband structure given by equation 4.25 is shown in Fig. 4.29(a) for a phase-modulated signal with modulation index $\alpha = 5$. There are eight significant sidebands on either side of the carrier so that the bandwidth varies in proportion to the modulating frequency. For the three modulating frequencies 5, 10 and 15 kHz, the bandwidths are 80, 160 and 240 kHz, respectively. Fig. 4.29(b) shows the situation for f.m. signals with the same modulating frequencies and with a peak frequency deviation of 75 kHz which is common in commercial broadcast work. The modulation index $\beta = D/\omega_m$ now varies with the modulating

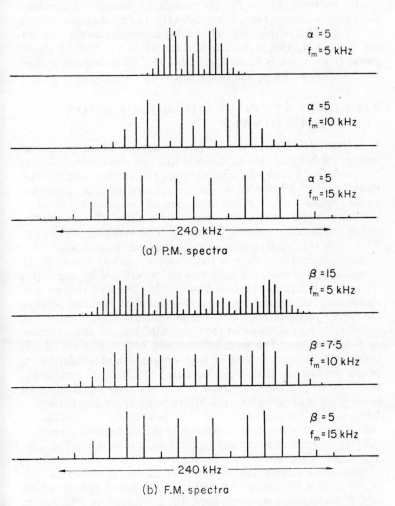

(a) P.M. spectra

(b) F.M. spectra

Fig. 4.29. Examples of sideband structure for (a) *p.m. signal with* $\alpha = 5$. (b) *f.m. signal with peak frequency deviation 75 kHz. Modulating frequencies in either case are 5, 10 and 15 kHz*

frequency from $\beta = 15$ at 5 kHz to $\beta = 5$ at 15 kHz. The bandwidths for the 5, 10 and 15 kHz modulating frequencies are found to be approximately 190, 220 and 240 kHz. The bandwidth occupied by the signal is in this case only weakly dependent on the modulating frequency. This system, which maintains $\beta \geqslant 5$ at all modulating frequencies, is termed *wideband f.m.* to distinguish it from *narrowband f.m.* for which $\beta \ll 1$, discussed earlier.

4.12 GENERATION AND DEMODULATION OF F.M. SIGNALS

It is possible to construct circuits or devices which present an impedance between two terminals whose reactive component may be varied by an externally applied voltage or current. The varactor diode is one such device whose capacitance may be controlled by an applied voltage. When made part of the tuned circuit of an oscillator, these devices may be used to allow the modulating signal to vary the oscillating frequency in the way required for f.m. This method of modulation is usually termed 'direct f.m.'.

As an alternative, there is the 'indirect' method due to E. H. Armstrong. The basic principle may be understood by comparing the phasor diagrams of Fig. 4.24 and 4.26. It is seen that an a.m. signal may be changed into a *narrowband* p.m. signal by rotating the carrier phase through 90°. The modulation depth of the a.m. signal must be small so that the locus ABC may be approximated as a circular arc. The modulation index is therefore necessarily small (so that a narrowband signal is obtained) and is independent of the modulating frequency. Note that the result is p.m. rather than f.m. In order to obtain f.m., the modulating signal is passed through an integrating network before being applied to the modulator, so that the sideband amplitudes (and hence the modulation index) become inversely proportional to the modulating frequency. A block diagram of an Armstrong modulator is shown in Fig. 4.30. The heart of the system is a balanced modulator which produces a suppressed carrier a.m. signal. The carrier is reinserted after being given a 90° phase shift and the final output is narrowband f.m. If wideband f.m. is required, the modulator is followed by frequency multipliers which raise both the carrier frequency and the peak frequency deviation (and hence β) by the same factor.

The reception of f.m. signals requires a demodulator or *discriminator* circuit whose output is proportional to the departure of the instantaneous signal frequency from the unmodulated carrier

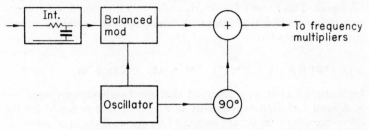

Fig. 4.30. Indirect (Armstrong) f.m. system

frequency. Ideally the discriminator should provide a linear response to frequency changes in the range $\omega_c \pm D$. It is also desirable for the discriminator to be completely insensitive to amplitude changes for, although the transmitted signal is of constant amplitude, the received signal may carry amplitude fluctuations due to interference from neighbouring signals or from motor car ignition. To prevent such amplitude changes from influencing the discriminator output, a voltage limiter is usually incorporated immediately following the i.f. amplifier (see Fig. 4.31). This clips the signal to a fixed level so that the final resonant circuit delivers a sine wave of constant amplitude to the discriminator. The limiter is essential if full advantage is to be taken of the relative immunity of wide-band f.m. to noise and interference (see below).

A discussion of discriminator circuits is beyond the scope of this text but a method which is becoming popular is to use a fairly low i.f. and to use this signal to trigger a pulse generator which delivers a pulse of fixed amplitude and width once per cycle of the

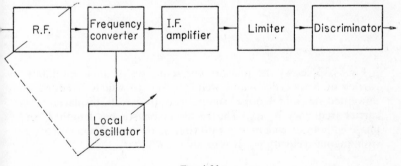

Fig. 4.31.

145

i.f. signal. This pulse train clearly contains a low-frequency component at the modulating frequency which may be extracted using a simple low-pass filter.

4.13 INTERFERENCE IN A.M. AND F.M.

In Section 4.11 it was observed that the bandwidth required for wideband f.m. signals is greatly in excess of that required for the corresponding a.m. or narrowband f.m. signal. Typical commercial f.m. sound broadcast transmissions, with a peak frequency deviation of 75 kHz, occupy a bandwidth of some 250 kHz. Transmission is usually in the relatively uncrowded v.h.f. (30–300 MHz) band but, even so, some justification is needed for this lavish use of bandwidth for the transmission audio signals, which before modulation occupy only 15 kHz or so. The justification lies in the superior signal-to-noise ratio enjoyed by f.m. systems, which means that for a given signal-to-noise ratio in the receiver output the transmitter power can be much smaller using f.m. than using a.m. To show how this comes about we shall, in this section, compare the interference produced in each case by an interfering sine wave on an unmodulated carrier. For simplicity it is assumed that the interfering signal is weak compared with the carrier.

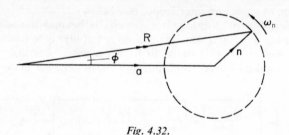

Fig. 4.32.

Fig. 4.32 shows the phasor representation of an unmodulated carrier of amplitude a and frequency ω_c to which is added an unwanted sinusoidal 'noise' signal, $n \cos (\omega_c + \omega_n)t$, displaced from carrier frequency by ω_n. The resultant phasor has length R and phase ϕ, both of which vary with time as the 'noise' phasor rotates with angular velocity ω_n. It is easily shown that

$$R = \sqrt{(a^2 + 2an \cos \omega_n t + n^2)} \qquad 4.26$$

and

$$\phi = \tan^{-1} \frac{n \sin \omega_n t}{a + n \cos \omega_n t} \qquad 4.27$$

We shall assume that $n \ll a$ so that, approximately,

$$R \approx a\left(1 + \frac{n}{a} \cos \omega_n t\right) \qquad 4.28$$

$$\phi \approx \frac{n}{a} \sin \omega_n t \qquad 4.29$$

An a.m. receiver will ignore the phase changes and interpret the signal as though it were modulated to a depth n/a by a sine wave of angular frequency ω_n. Let us suppose the receiver sensitivity is such that 100% modulation gives a 1 V peak signal from the demodulator. The interfering sine wave will thus have an amplitude n/a or power

$$P_{\text{a.m.}} = \frac{n^2}{2a^2} \qquad 4.30$$

On the other hand, an f.m. receiver will ignore the amplitude changes and sense an instantaneous frequency deviation given by

$$\frac{d\phi}{dt} = \frac{n\omega_n}{a} \cos \omega_n t$$

To obtain a fair comparison with the a.m. receiver we will suppose the discriminator is such that a frequency deviation equal to the peak frequency deviation D, chosen in the design of the f.m. system, gives an output of 1 V. The interference thus gives a sine wave output at frequency ω_n and $n\omega_n/aD$ peak. The corresponding power is

$$P_{\text{f.m.}} = \frac{n^2 \omega_n^2}{2a^2 D^2} \qquad 4.31$$

The results expressed in equations 4.30 and 4.31 are illustrated in Fig. 4.33(a) and (b). In both cases the interference will be stopped by the audio amplifier stages when $\omega_n > \omega_b$, where ω_b is the highest modulating frequency expected ($\omega_b = 2\pi \times 15\,000$, say). It follows that the f.m. receiver will always be superior to the a.m.

147

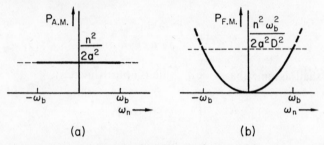

Fig. 4.33. Power output from the demodulator of a.m. and f.m. receivers arising from an interfering sinusoid displaced ω_n from carrier

one, provided

$$\frac{n^2\omega_b^2}{2a^2D^2} < \frac{n^2}{2a^2}$$

or

$$\frac{\omega_b}{D} < 1$$

i.e.

$$\frac{D}{\omega_b} = \beta_b > 1$$

where β_b is the modulation index corresponding to the *highest* modulating frequency used. In practical wideband f.m. systems $\beta_b \approx 5$ so that a substantial improvement is achieved. The initial assumption, $n \ll a$, should be borne in mind, however. The interference in the output from a wideband f.m. receiver is less than that of a corresponding a.m. receiver only when the *incoming* signal-to-noise ratio is high. The same result applies with interference which is uniformly distributed throughout the i.f. bandwidth instead of being localised at one particular frequency (see Problem 4 and Section 7.9).

The situation when $n > a$ is illustrated in Fig. 4.34. It is seen that the locus of the end point of the resultant phasor now completely encircles the origin. This resultant is best regarded as a signal of frequency $\omega_c + \omega_n$ carrying 'interference' due to the signal $a \cos \omega_c t$. When $n < a$, as in Fig. 4.32, the maximum phase shift introduced by the interfering phasor cannot exceed $\pi/2$. If the carrier a is modulated, then phase shifts $\beta \gg \pi/2$ are produced by the modulation. The additional phase shifts by the interfering sinusoid are then relatively small. On the other hand, when $n > a$, as in Fig. 4.35, the noise 'takes over' and produces very large phase

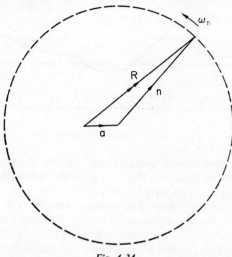

Fig. 4.34.

changes. If the interfering signal is a modulated f.m. carrier then the discriminator will be dominated by the interfering signal. This so-called 'capture effect' in f.m. systems means that when two modulated carriers are present in the i.f. bandwidth the large signal tends to dominate the smaller, even though the receiver may be tuned to the smaller one.

4.14 PULSE AMPLITUDE MODULATION

In the modulation systems discussed so far, the message waveform has been made to vary one of the parameters—amplitude, phase, or frequency—of a sinusoidal carrier. Instead of a sine wave carrier it may be preferred, in certain applications, to use a train of pulses. The message signal is then arranged to modulate some parameter of the pulse train, such as the pulse amplitude, repetition frequency, or pulse width. Of the many possible varieties of pulse modulation, only three will be examined here. First, pulse amplitude modulation (p.a.m.), which is the pulse equivalent of sine wave carrier a.m. Second, pulse position modulation (p.p.m.), which is similar to the phase modulation (or frequency modulation) of a sine wave carrier. The third example, pulse code modulation (p.c.m.), is unique to pulse train carriers and has no sine wave equivalent.

149

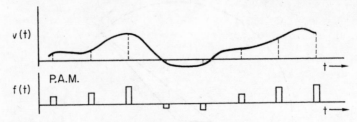

Fig. 4.35. Pulse amplitude modulation

All pulse modulation systems rely on the sampling theorem (Section 2.15), which shows that it is possible to describe a continuously varying message waveform in terms of a sequence of statements of its sample values. The message waveform must be band-limited, and if the upper frequency limit is F Hz then samples must be taken at least $2F$ times per second. Fig. 4.35 illustrates the process of p.a.m. in which the sample values, shown dotted, are used to control the amplitudes of successive pulses in the pulse train. The pulses have been sketched as rectangular pulses. One method of producing a p.a.m. signal, however, would be to gate the modulating waveform by means of a train of pulses, in which case the tops of the pulses would not be flat but would have the shape of the short section of the waveform from which they were taken. If the pulse width is small compared with the repetition interval the difference may be taken as negligible, so that in either case the spectrum of the modulated p.a.m. train is essentially that of Fig. 3.32. This shows that the original waveform may be recovered by passing the modulated pulse train through a low-pass filter with cut-off frequency F.

The p.a.m signal of Fig. 4.35 occupies a wide bandwidth and this is somewhat wasteful unless advantage is taken of the possibility of multiplexing. Fig. 4.36 illustrates *time division multiplexing* for three channels, each band-limited to F Hz. A timing pulse generator supplies $6F$ pulses per second to a sampling unit which samples channels 1, 2, and 3 sequentially. The outcoming t d.m. signal is sketched in Fig. 4.37(a). Pulses 1, 1, 1, ... form a p.a.m. train corresponding to channel 1. Interleaved with these are the pulses 2, 2, 2, ... and 3, 3, 3, ... which belong to channels 2 and 3, respectively. When this pulse train is received, the pulses may be rerouted sequentially into three low-pass filters whose outputs will be the three separate message waveforms once more. By this

150

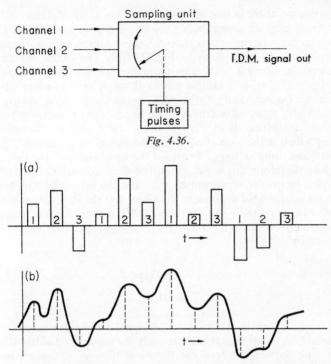

Fig. 4.36.

Fig. 4.37. (a) *A time-division-multiplexed p.a.m. train* (b) *The same signal after passing through a channel of finite bandwidth*

means, a single communication link may be used to transmit several messages simultaneously.

What is the minimum bandwidth required for a time-division-multiplexed p.a.m. signal? When the signal of Fig. 4.37(a) is passed through a system having finite bandwidth, the resulting amplitude and phase distortion will lead to a received waveform which might look rather like Fig. 4.37(b). This signal may now be sampled, in correct synchronism, to yield the separate p.a.m. trains, but there is a danger that the overshoot and ringing, due to distortion on the link acting on a given outgoing pulse, will not have died away before the next sample is taken. This means that a given sample may suffer interference *(intersymbol distortion)* from preceding pulses in the train, leading to interference between one channel and another *(cross-talk)*.

11

151

In theory, there is one idealised situation in which intersymbol distortion may be completely avoided. Suppose that, instead of using rectangular pulses to transmit these samples, sinc x-shaped pulses are used instead, as in Fig. 2.17(b). The shape chosen is sinc $(\pi t/T)$, where T is the sampling interval, for then the pulse corresponding to one sample passes through zero at every other sampling instant. Only three such pulses have been drawn in Fig. 2.17(b), to avoid confusion. The bandwidth of each sinc pulse is, from transform B of Fig. 2.14, $\omega_c = \pi/T$ and the composite signal which is the sum of all these pulses will be a smooth waveform band limited to ω_c. Provided the bandwidth of the link exceeds ω_c therefore, this waveform will be received undistorted and if sampled in proper synchronism will yield the original t.d.m. train with no intersymbol distortion. Recalling that the sampling interval T must be equal to or shorter than $1/2MF$ where M is the number of channels being multiplexed, the minimum channel bandwidth is given by

$$\omega_c = \pi/T$$
$$= 2\pi MF$$
$$= M\omega_m$$

where $\omega_m = 2\pi F$ is the modulating signal bandwidth in rad s^{-1}. It should be noted that this is precisely the same bandwidth which would be required to transmit the same number of signals by a frequency-multiplexed system.

In practice, of course, sinc $\omega_c t$-shaped pulses cannot be used and some approximation to band-limited pulses would be made by choosing a rounded pulse, possibly of the cosine-squared form, rather than sharp rectangular pulses. In order to make synchronous sampling possible at the receiver, it would also be necessary to transmit timing pulses once for every complete sequence of M channel pulses, so that $\omega_c = (M+1)\omega_m$ would be a more realistic estimate of the minimum required bandwidth.

4.15 OTHER FORMS OF PULSE MODULATION

If the channel carrying a p.a.m. signal is subject to noise and interference, then the added noise will change the pulse amplitudes just as though the noise had been added to the message waveform prior to modulation and nothing can be done to eliminate it from the received signal. Pulse position modulation, on the other hand, may be made less susceptible to added noise. In this system the waveform is again sampled at the Nyquist rate but the samples are used to displace the carrier pulses from their unmodulated positions. These displacements are made proportional to the sample values as shown in Fig. 4.38. The receiver generates pulses whose amplitudes are proportional to the time delay between pulses from a synchronous local pulse train and the leading edges of the pulses received. The resulting p.a.m. train is passed through a low-pass filter to complete the demodulation.

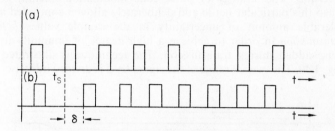

Fig. 4.38. Pulse position modulation (p.p.m.)
(a) *Unmodulated carrier*   (b) *Modulated pulse train.*
The delay is made proportional to the signal sampled at t_s

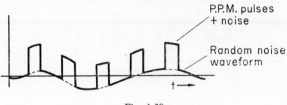

Fig. 4.39.

When noise is present with the p.p.m. signal the pulses will be displaced up or down, as shown in Fig. 4.39. The position of the leading edge—which for simplicity is assumed to be vertical—is not changed, however, by the added noise. There will thus be no interference on the demodulated waveform. In practice, of course, the transmission bandwidth must be finite and each vertical step becomes one with a finite rise time. The pulse position must then be judged as the moment at which the signal level reaches some prescribed value—say one-half the pulse height. In this case some

11*

noise will appear on the received waveform but, as will be argued in Section 7.10, a significant improvement over p.a.m. is possible if the rise times are kept short. A short rise time implies a wider transmission bandwidth than would be needed for p.a.m. We have here another example of 'wideband improvement' in signal-to-noise ratio, similar to the improvement of wideband f.m. over a.m.

As we shall see in Chapters 6 and 7, added noise cannot be completely avoided in any system and it is inevitable that the sample values of a message waveform cannot be transmitted with perfect precision. The strategy of pulse code modulation (p.c.m.) is to grasp this particular nettle and deliberately allow a controlled and tolerable amount of uncertainty in the sample values *before transmission* in order to achieve a high degree of immunity from noise added during transmission. The technique is illustrated in Fig. 4.40. The overall range of the signal level is divided into discrete

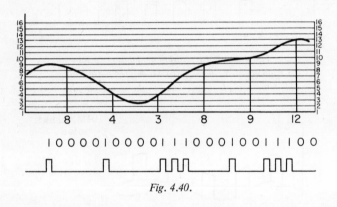

Fig. 4.40.

steps, represented in this example by the voltage levels labelled 1 to 16. The sample values, shown as vertical lines, are not measured exactly but only to the next level below—a procedure called *quantisation*. The first row of numbers beneath the graph gives the quantised sample values in decimal notation. In practice binary notation is preferred and the corresponding binary code words have also been written without a break at the bottom of the figure. In this way the continuously varying waveform has been coded into a sequence of binary digits, each group of four digits representing a sample value. The electrical analogue, to be transmitted, is the pulse train shown.

154

The primary function of the receiver is to sample the received waveform (+ noise) in proper synchronism and to decide at each moment whether a pulse is present or not. Now, even if a pulse has been seriously changed in amplitude and shape by noise and distortion in the communication link, the choice between 1 and 0 may still be made with a high degree of certainty. It will be shown in Section 7.11 that virtually error-free transmission can be achieved at signal-to-noise ratios which would be intolerable with p.a.m. This is not to say that the system is noise-free, for even when error-free binary transmission is accomplished there remain the sampling errors which are inherent in the quantisation process. The effect of these errors—*quantisation noise*—may be reduced by using more closely spaced quantisation levels, but this will require longer binary code words and therefore a faster pulse repetition rate, i.e. a wider bandwidth. Once more, the way to combat noise is seen to be by increasing the bandwidth.

The bandwidth required for p.c.m. may be found as follows. Let there be L quantisation levels. This will require $\log_2 L$ binary digits for each sample value, and the sample values must be taken $2F$ times per second if the signal band limit is F Hz. The transmitted pulse train will therefore contain

$$n = 2F \log_2 L$$

pulses (amplitudes 1 or 0) per second. The sampling theorem states that n samples per second may be transmitted using a bandwidth $n/2$ Hz. Thus the minimum bandwidth, f_c, required for the p.c.m. channel is

$$f_c = F \log_2 L$$

which expresses the numerical relationship between the number of levels (and thus the quantisation noise) and the bandwidth.

A block diagram of a single-channel p.c.m. link is sketched in Fig. 4.41. The message waveform is first sampled to yield a p.a.m. signal and this is passed to an analogue-to-digital converter which generates a binary code word corresponding to the nearest quantisation level. At the receiver the procedure must be reversed using a digital-to-analogue converter. Time division multiplexing is usually arranged by interleaving the code words of the separate signals, rather than by allocating alternate binary digits to separate channels, for then a single analogue-to-digital converter may be used for the whole multiplexed signal. The complexity of the electronic equipment has meant that, although the merits of p.c.m.

155

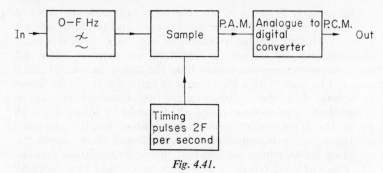

Fig. 4.41.

have been appreciated for many years, its commercial exploitation has been delayed until the comparitively recent introduction of transistor and integrated-circuit techniques. Further details of this most elegant modulation system may be found in references 5, 6, and 7.

References

1. ROCHE, A. H., and WEAVER, L. E., 'Television frequency—translating terminal equipment for the Birmingham–Holme Moss Coaxial Cable', *Proc. Instn elect. Engrs* **99**, 3A, 455 (1952)
2. TAYLOR, F. J. D., 'Carrier system No. 7', *P.O. elect. Engrs' J.* **34**, 101, (1941)
3. WOOSTER, C. B., 'An introduction to telephone channelling', *Point to Point Telecomm.*, **3**, No. 2, 5 (1959)
4. BRAY, W. J., and MORRIS, D. W., 'Single sideband multi-channel operation of short-wave point-to-point radio links', *P.O. elect. Engrs' J.*, **45**, 97 (1952)
5. OLIVER, B. M., PIERCE, J. R., and SHANNON, C. E., 'The philosophy of p.c.m.', *Proc. Inst. Radio Engrs* **36**, 1324 (1948)
6. WHYTE, J. S., 'Pulse code modulation', *P.O. elect. Engrs' J.* **54**, 86 (1961)
7. STEVENS, A. D., 'A p.c.m. system for junction telephone circuits', *Point to Point Telecomm.*, **10**, No. 2, 6, and No. 3, 40 (1966)

Further Reading

BLACK, H. S., *Modulation Theory* (Van Nostrand, New York, 1953)
TIBBS, C. E., and JOHNSTONE, G. G., *Frequency Modulation Engineering* (Chapman and Hall, London, 1956)
BETTS, J. A., *High Frequency Communication* (E.U.P., London, 1967)
BETTS, J. A., *Signal processing, modulation and noise* (E.U.P., London, 1970)

PROBLEMS

1. An unmodulated carrier has mean power P_0. When it is amplitude-modulated by a sinusoid the power becomes P. Show that

156

the modulation depth is

$$m = \sqrt{\left[\frac{2(P-P_0)}{P_0}\right]}$$

2. Sketch the form of the output from an envelope detector when the following signals are applied.

    (i) An r.f. carrier, amplitude modulated to a depth of 10% by a sinusoid.

    (ii) The same signal with one sideband removed.

    (iii) The signal of (i) with the carrier removed.

3. A carrier of amplitude a is sinusoidally modulated to a depth m. The lower sideband is removed and the resulting signal applied to an envelope detector. Show that the output is given by equation 4.26 with $n = am/2$. From a binomial expansion which includes terms up to m^2, show that the output may be written

$$R = a\left(1 + \frac{m^2}{16} + \frac{m}{2}\cos \omega_m t - \frac{m^2}{16}\cos 2\omega_m t + \ldots\right)$$

Find the percentage second harmonic distortion when $m = 10\%$.

4. Consider the response of an envelope detector to a.m. signals which have been passed through either of the filters whose characteristics (neglecting phase distortion) are sketched here, for modulating frequencies both greater and smaller than f_v.

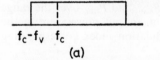

(a)

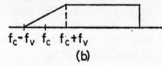

(b)

Fig. 4.42.

(An a.m. signal filtered as in (a) in the sketch is used in television broadcasting in order to conserve bandwidth. At the receiver the spectrum is shaped as in (b). This system is termed vestigial sideband transmission. In 625-line transmissions in Britain the vestige width $f_v = 1\cdot25$ MHz, saving $4\cdot25$ MHz on the full sideband width of $5\cdot5$ MHz.)

5. Three receivers are tuned to the same unmodulated carrier. Receiver A is an a.m. receiver with an i.f. bandwidth of 8 kHz and employs an envelope detector. Receiver B is an f.m. receiver with an i.f. bandwidth of 200 kHz and a discriminator designed for a peak frequency deviation of 75 kHz. Receiver C is a p.m. receiver with the same i.f. bandwidth as B and a demodulator designed for a maximum phase shift of 5 radians. All the receivers have audio stages which cut off at 6 kHz. In each case describe what will be heard when an interfering (unmodulated) carrier, 20 dB below the desired carrier at the input terminals, is swept in frequency from 50 kHz below to 50 kHz above the frequency to which the receivers are tuned.

6. A 1 MHz carrier is frequency-modulated by a 1 kHz sinusoid to a peak frequency deviation of ± 3 kHz. Determine the bandwidth required. What bandwidth is required when the modulating signal is reduced in amplitude by 50%?

7. Show that a d.s.b. signal may be multiplexed to carry two modulating signals by using two (suppressed) carriers in phase quadrature.

8. A 2 kHz carrier is frequency-modulated by a 100 Hz square wave so that the carrier steps between 1·7 kHz and 2·3 kHz. By considering this as the superposition of two a.m. waves, find the spectrum and the bandwidth required.

9. Fig. 4.34 gives the power of the output signals from a.m. and f.m. demodulators, for interference whose power is concentrated at ω_n. If many (small-amplitude) 'noise' sinusoids are distributed along the frequency axis, the total output powers may be compared by integrating $P_{\text{a.m.}}$ and $P_{\text{f.m.}}$ over the range $\pm \omega_b$. Show that the a.m. and f.m. output powers are in the ratio $3\beta_m^2 : 1$, where β_m is the modulation index corresponding to modulation at ω_b.

10. A p.a.m. signal, composed of short sample pulses occurring at $2F$ per second, is demodulated by passing it through a low-pass filter with cut-off frequency F. Show that the filter reconstructs the original waveform in the way expressed mathematically by equation 2.52.

Chapter 5

Probability and Statistics

5.1 INTRODUCTION

The voltage or current observed at the receiving end of a communication channel is unpredictable, at least between certain limits. The message signal cannot be predicted, for if it could, then no information would be gained by observing it; a predictable signal would not be 'news'. Moreover the received signal will be perturbed by random fluctuations, or 'noise', arising in the channel or in the receiver, whose effect is to increase our uncertainty about the transmitted message. In general, however, certain average properties will be discernible in both message and noise. For example, in a teleprinter signal the regularity of occurrence of the letter E and also the received noise power will have fairly well-defined average values.

Signals which carry information thus belong to that class of natural phenomena which must be regarded as random and unpredictable, either because detailed calculations become impossibly complex—as in the case of a large number of interacting gas molecules—or because the causal agents are unknown—as in the rate of incidence of certain diseases—or because of the inherent uncertainty asserted by wave mechanics. In such cases, although the detailed pattern of behaviour cannot be predicted, certain statistical regularities exist and may be treated mathematically through the theory of probability. Statistical analysis plays an important role in communication theory, particularly as regards understanding and

159

combating the deleterious effects of noise. In this chapter certain
rudimentary concepts in the theory of probability will be outlined
in order to provide material which will be needed later. For a more
rigorous and balanced account, however, the reader is referred to
the texts devoted to the subject, a few of which are listed at the end
of the chapter.

5.2 PROBABILITY OF AN EVENT

We begin by considering some experiment or trial whose possible
outcomes may be classified, as finely as seems necessary, into a
finite set of all possible outcomes. These outcomes will be referred
to as *events* where an event is simply something which may or may
not happen as a result of a trial. It is assumed that the events of a
given classification are mutually exclusive so that if we are told that
some event has occurred then no other event belonging to that
classification can possibly have occurred simultaneously. For
example, consider a trial consisting of the random selection of a
coin from a large number of coins. In this case the events might be
classified according to the denomination of the coin selected. Or,
again, if a die is thrown the events might be represented by the
integers 1 to 6.

Suppose that the possible outcomes of an experiment have been
divided into n mutually exclusive events E_k. We may then assign
a number $P(E_k)$ to each event which we will call the *probability of
the event*. The numbers $P(E_k)$ are made subject to the conditions

$$\left.\begin{array}{c} 0 \leqslant P(E_k) \leqslant 1 \\ \sum_{k=1}^{n} P(E_k) = 1 \end{array}\right\} \qquad 5.1$$

If an event is certain to occur then we give it unit probability but,
beyond this, the precise allocation of the numbers $P(E_k)$ is left open
for the present. Sometimes, as in the case of a 'fair' die, we may
appeal to symmetry and attach equal probabilities to each event,
but the ultimate test lies in whether or not the assigned probabilities together with the axioms of the theory provide a description of
the random phenomenon which accords with experience. We note,
however, that the 'law of large numbers' equates $P(E_k)$ with the
relative frequency of occurrence of the event E_k in an indefinitely
large number of trials (see Section 5.11 below) and also that the

160

axiomatic assumptions such as 5.1 fall naturally into place if probabilities are assigned on a relative frequency basis.

If E_j and E_k are two mutually exclusive events then we define the probability of the event (E_j OR E_k) to be

$$P(E_j \text{ OR } E_k) = P(E_j) + P(E_k) \qquad 5.2$$

This fundamental rule will be referred to as the 'additive law'. It applies only to events which are mutually exclusive.

5.3 CONDITIONAL PROBABILITY AND JOINT PROBABILITY

It is often possible to classify the possible outcomes of a trial in two (or more) different ways so that there is a set of mutually exclusive events A_j and also a set of mutually exclusive events B_k. Each outcome will then correspond to both an A_j and a B_k. For example, if a person is selected at random from a given group of students the events A_1 and A_2 may correspond to 'the person selected is male' and 'the person selected is female', respectively. Alternatively, we may make a distinction according to dark, fair, or red hair and denote these events by B_1, B_2 and B_3. We may assign probabilities to these events such that

$$P(A_1) + P(A_2) = 1$$

and

$$P(B_1) + P(B_2) + P(B_3) = 1$$

Now consider the question, 'What is the probability that the person selected has red hair, given that he is male?' This question asks for a *conditional probability* and we use the notation $P(B_3 | A_1)$ which may be read, 'the probability of event B_3 given that A_1 occurred'. Similarly, the probability that the person is female if the hair colour is dark would be denoted $P(A_2 | B_1)$. In statements of conditional probability, the event which is thought of as having occurred is commonly termed the *hypothesis*.

Alternatively we may ask the question, 'What is the chance that *both* events A_j and B_k will occur?' We will denote this *joint probability* as $P(A_j, B_k)$. Thus, in the example above, $P(A_2, B_2)$ is the chance of selecting a blonde girl.

Conditional and joint probabilities will also arise in the discussion of experiments consisting of two parts. The noise voltage across a resistor may be measured at time t_1 and then again at

161

time t_2, these two measurements being regarded as a single experiment. In this case we may take the events A to correspond to the voltage at t_1 and the events B to correspond to that at t_2. Similarly, in a trial consisting of two throws of a die, the events A would be the possible results of the first throw and the events B those of the second throw.

We now seek relationships between the elementary probabilities $P(A_j)$, $P(B_k)$, the conditional probabilities $P(A_j|B_k)$, $P(B_k|A_j)$ and the joint probabilites $P(A_j, B_k)$. Suppose there are n events A_j and m events B_k; then there $n \times m$ joint events (A_j and B_k). Moreover, the joint events are mutually exclusive so that it follows from the additive law that

$$P(A_j) = \sum_{k=1}^{m} P(A_j, B_k)$$

and

$$P(B_k) = \sum_{j=1}^{n} P(A_j, B_k)$$

5.3

In order to develop a rule for calculating the conditional probabilities in terms of the joint probabilities let us return to our example and suppose that 56% of the student group are females $(P(A_2) = 0.56)$ and 7% of the group are red-haired females $(P(A_2, B_3) = 0.07)$. If we are now told that the person selected is female then the chance that she has red hair would be $P(B_3|A_2) = 7/56 = 0.125$. We generalise this and define

$$P(B_k|A_j) = \frac{P(A_j, B_k)}{P(A_j)}$$

and

$$P(A_j|B_k) = \frac{P(A_j, B_k)}{P(B_k)}$$

5.4

Combining equations 5.3 and 5.4 it follows that

$$P(B_k|A_j) = \frac{P(B_k)\, P(A_j|B_k)}{\sum_{k} P(B_k)\, P(A_j|B_k)}$$

and

$$P(A_j|B_k) = \frac{P(A_j)\, P(B_k|A_j)}{\sum_{j} P(A_j)\, P(B_k|A_j)}$$

5.5

Equations 5.5, usually known as Bayes's formulae, enable us to evaluate one set of conditional probabilities given the other set and the corresponding elementary probabilities.

5.4 STATISTICAL INDEPENDENCE

The conditional and joint probability relationships for two events A and B may be written, following equations 5.4, in the form

$$P(A, B) = P(A)P(B\,|\,A) = P(B)P(A\,|\,B) \qquad 5.6$$

Now let us suppose that the knowledge that A has occurred leaves the probability of B unchanged, i.e.

$$P(B\,|\,A) = P(B)$$

In this case the second equality in 5.6 reads

$$P(A)P(B) = P(B)P(A\,|\,B)$$

so that

$$P(A\,|\,B) = P(A)$$

Thus the converse is also true: the knowledge that B has occurred leaves the probability of A unchanged. Under these circumstances equation 5.6 becomes

$$P(A, B) = P(A)P(B) \qquad 5.7$$

This equation expresses the *product law* for independent events. It may be extended to cover more than two events.

We often make the assumption that trials which are repeated under identical circumstances yield a sequence of events which are statistically independent. For example, if a coin is tossed three times, the probability of obtaining the result 'head' at each toss is $\frac{1}{2}$ so that the probability of obtaining the sequence 'head, head, head' is $\frac{1}{2} \times \frac{1}{2} \times \frac{1}{2} = \frac{1}{8}$. The probability of any other sequence of three is also $\frac{1}{8}$. It is important to appreciate that, no matter how long the sequence, the probability of obtaining a result such as

$$H\,H\,H\,H\,H\,H\,H\,H\,H$$

is the same as that for any other *particular* sequence, say

$$H\,H\,T\,H\,T\,T\,H\,H\,T\,T$$

although the latter may appear more likely. The crucial point is that we are assuming statistical independence so that a sequence of nine heads leaves the probability of getting a head at the next toss unchanged at $\frac{1}{2}$.

5.5 RANDOM VARIABLES, PROBABILITY FUNCTIONS, MEAN, AND VARIANCE

In many important problems concerning random phenomena the observed result of a trial is simply a number. We then say that we are observing a *random variable* and that each trial yields one value of the random variable. If the random variable is denoted by X, then to each event there corresponds a point x on the axis of real numbers. For example, the trial may be the observation of a thermometer at a particular location at noon on a certain day of the year, in which case the event 'the mercury thread extended to this particular mark' is recorded as a number which is the value of the random variable 'temperature'. Throwing dice provides another example and here the appearance of a certain pattern of spots leads us to record an integer. Now, even if the events themselves do not naturally give rise to numbers, we may choose, if it seems useful, to define some rule which attaches a number to each event. In spinning a coin the outcome 'heads' may be recorded as 1 and 'tails' as 0. To a sequence of 10 symbols received from a teleprinter we may attach an integer equal to the number of letter A's in the sequence. In this last example we note that several different events may give rise to the same number, for many different words contain the same number of A's. Again, if two dice are thrown, we may only be interested in the total number of spots, so that the events (4,4), (5,3), (6,2), etc., each 'map' to the same integer, 8. In this case the rule (or function) 'sum of the numbers of spots' is the random variable. The language of mathematics would describe a random variable as a function whose domain is the set of events and whose range is part or all of the axis of real numbers. Thus in the teleprinter example above, the given function operating on the event 'bananapeel' takes the value 3. In the material to follow we will use a small letter, for example x, to denote a particular value which the random variable may take, in order to make a clear distinction between this and the random variable itself, which will be denoted by the corresponding capital letter, X.

If the range of a random variable X includes only a finite number, n, of values x_j then it is said to be *discrete*. Each value x_j has a probability of occurrence $P(x_j)$ and the function $P(x_j)$ is called the probability function of the random variable X. It is clear that equation 5.2 requires

$$\sum_{j=1}^{n} P(x_j) = 1 \qquad\qquad 5.8$$

A probability function is characterised by certain quantities which we now define. First the *mean* or *average* or *expectation* of X is given by

$$\mu_X = \sum_j x_j P(x_j) \qquad 5.9$$

and may also be designated $E(X)$ or $\bar{X}$. Similarly the mean square of X is written

$$E(X^2) = \sum_j x_j^2 P(x_j) \qquad 5.10$$

It is sometimes helpful to think of the $P(x_j)$ as masses or weights along the X-axis. With this picture in mind the mean μ_X is often termed the 'first moment' of the probability function since it is the result of taking moments about the origin. The mean square $E(X^2)$ is analogous to the moment of inertia taken about the origin. If on the other hand we take the second moment (moment of inertia) about the mean (centre of gravity) we obtain the quantity

$$\sigma_X^2 = \sum_j (x_j - \mu_X)^2 P(x_j) = E[(X - \mu_X)^2] \qquad 5.11$$

The quantity σ_X^2, also written Var (X), is termed the *variance* of X. The square root of the variance, σ_X, is the *standard deviation*.

A useful relationship is obtained upon expansion of 5.11 and using 5.9 and 5.10

$$\sigma_X^2 = E(X^2) - \mu_X^2 \qquad 5.12$$

which corresponds to the parallel axes theorem in dynamics.

5.6 CHEBYSHEV'S INEQUALITY

The variance of a random variable may be thought of as a measure of the spread of its probability function about the mean. The smaller σ_X^2 the less likely to occur are values of X lying far from μ_X. More precisely, the probability that a value x will occur outside the range $\mu_X \pm c$, where c is arbitrary, may be given a definite upper limit if the variance is known. To demonstrate this, we begin with equation 5.11 and note that as every term in the summation is non-negative we may write

$$\sigma_X^2 \geqslant \sum (x_j - \mu_X)^2 P(x_j)$$

where the summation now extends only over those x_j for which

$(x_j - \mu_X)^2 \geqslant c^2$. The inequality can only be increased if we now substitute c^2 for $(x_j - \mu_X)^2$ in this last equation, so that

$$\sigma_X^2 \geqslant c^2 \sum P(x_j)$$

The restricted sum here adds the probabilities for all x_j for which $|x_j - \mu_X| > c$ and thus represents the probability that X will take a value lying outside the range $\mu_X \pm c$. Thus we may write

$$P(|X - \mu_X| > c) \leqslant \frac{\sigma_X^2}{c^2} \qquad 5.13$$

a result known as Chebyshev's inequality.

5.7 PROBABILITY DENSITY FUNCTIONS

When the range of a random variable is some continuous sequence of numbers, which may be the whole of the axis of real numbers, then it is clear that the probability of occurrence of some precise value, x, is zero. We may, however, speak of the probability that the variable will take a value lying in some interval between x and $x + \delta x$. This probability may be divided by δx to yield a quantity having the characteristics of a density, and in the limit, as $\delta x \rightarrow 0$, becomes the probability density function $p(x)$. The probability that X will take a value in the range $a \leqslant X \leqslant b$ is thus given by

$$P(a \leqslant X \leqslant b) = \int_a^b p(x)\,\mathrm{d}x$$

The results obtained for discrete variables may be carried over in a straightforward manner when the random variable is continuous. Thus the requirement of normalisation expressed by 5.8 becomes

$$\int p(x)\,\mathrm{d}x = 1 \qquad 5.14$$

while the mean and variance are given by

$$\mu_X = \int x\,p(x)\,\mathrm{d}x \qquad 5.15$$

$$\sigma_X^2 = \int (x - \mu_X)^2\,p(x)\,\mathrm{d}x \qquad 5.16$$

where the range of integration in the last three equations is assumed to embrace the whole region of the x-axis where $p(x)$ is finite.

166

5.8 TWO RANDOM VARIABLES

Problems frequently arise in which each event yields the values of two (or more) random variables. For example, a person may be selected at random from a given population and asked to state both his height and his weight. One may ask for the amplitude and the phase of a particular Fourier component of a signal.

If the random variables are discrete there will be a joint probability $P(x_j, y_k)$ associated with each pair of values x_j and y_k as well as the elementary probabilities $P(x_j)$ and $P(y_k)$. In accordance with equations 5.3 we may write

$$P(x_j) = \sum_k P(x_j, y_k)$$
$$P(y_k) = \sum_j P(x_j, y_k)$$

5.17

The joint probabilities may be imagined as set down in a table with a row for each x_j and a column for each y_k. The value of $P(x_j)$ is then found by adding the numbers along the jth row and the result would be conveniently set down in the right-hand margin. Similarly, the sums down the columns, the $P(y_k)$, would be written along the lower margin. This kind of display gives rise to the term 'marginal probabilities' for $P(x_j)$ and $P(y_k)$.

Related to the joint probabilities are the conditional probabilities $P(x_j|y_k)$ and $P(y_k|x_j)$. To an observer who has access only to the numbers x_j and y_k as they occur, these values of the random variables form events A_j, B_k of the type discussed in Section 5.3 and the results expressed by equations 5.3, 5.4 and 5.5 are immediately available. In particular, following the argument of Section 5.4, we define two random variables to be statistically independent if

$$P(x_j, y_k) = P(x_j) P(y_k)$$

for *every* pair of values x_j and y_k. With the tabular display of the last paragraph in mind, this means that every entry $P(x_j, y_k)$ must equal the product of the corresponding marginal probabilities.

If X and Y are continuous then the marginal probabilities become probability density functions and we may also define a joint probability density function $p(x, y)$. Particular pairs of values x, y may be thought of as points in a Cartesian reference frame (which replaces the tabular display for discrete variables). The function $p(x, y)$

Signals and Information

now has the characteristics of a surface density. Thus the probability that, as a result of a trial, X takes a value between x_1 and x_2 while Y takes a value between y_1 and y_2 is given by

$$P(x_1 \leqslant X \leqslant x_2 \text{ and } y_1 \leqslant Y \leqslant y_2) = \int_{x_1}^{x_2} \int_{y_1}^{y_2} p(x, y) \, dx \, dy$$

The conditional probability density $p(x/y)$ is defined by

$$p(x \mid y) = \frac{p(x, y)}{p(y)}$$

and if the random variables are statistically independent we may write

$$p(x, y) = p(x) p(y) \qquad 5.18$$

5.9 SUMS AND PRODUCTS OF RANDOM VARIABLES

If two random variables X and Y appear as the result of a certain experiment then we may be interested in forming their sum or product. The sum $X + Y$ is in fact a new random variable which possesses a mean and a variance which we now wish to evaluate in terms of the means and variances of X and Y separately. For brevity we restrict ourselves to the case where X and Y are discrete; the results obtained will be true for continuous variables also. The expectation of $X + Y$ is, by definition, given by

$$\begin{aligned}
E(X+Y) &= \sum_{j,k} (x_j + y_k) P(x_j, y_k) \\
&= \sum_{j,k} x_j P(x_j, y_k) + \sum_{j,k} y_k P(x_j, y_k) \\
&= \sum_j x_j P(x_j) + \sum_k y_k P(y_k) \\
&= E(X) + E(Y) = \mu_X + \mu_Y \qquad 5.19
\end{aligned}$$

Thus the mean of the sum is the sum of the means. In the second step of this argument equations 5.17 have been used.

A simple extension of this theorem to the random variable $S = aX + bY$ gives the result

$$\mu_s = E(aX + bY) = a E(X) + b E(Y) = a\mu_X + b\mu_Y$$

168

The mean of the product of two random variables follows again from definition

$$E(XY) = \sum_{j,k} x_j y_k P(x_j y_k)$$

This expression simplifies if X and Y are independent, for then $P(x_j, y_k) = P(x_j) P(y_k)$ and hence

$$E(XY) = \sum_{j,k} x_j P(x_j) y_k P(y_k)$$

$$= \left[\sum_j x_j P(x_j) \right] \times \left[\sum_k y_k P(y_k) \right]$$

$$= E(X).E(Y) = \mu_X \mu_Y \qquad 5.20$$

So that for independent variables the mean of the product is the product of the means.

We are now in a position to find the variance of the sum $X+Y$. By definition (equation 5.11)

$$\begin{aligned}\text{Var}\,(X+Y) &= E\{[(X+Y)-E(X+Y)]^2\}\\ &= E\{[(X-\mu_X)+(Y-\mu_Y)]^2\}\\ &= E[(X-\mu_X)^2]+E[(X-\mu_Y)^2]\\ &\quad +2E[(X-\mu_X)(Y-\mu_Y)] \qquad 5.21\end{aligned}$$

In arriving at 5.21 we have used an extension of 5.19 to the three random variables $(X-\mu_X)^2$, $(Y-\mu_Y)^2$ and $2(X-\mu_X)(Y-\mu_Y)$. The first and second terms in 5.21 are σ_X^2 and σ_Y^2; the third term is easily shown to be zero upon expansion and application of 5.20, provided that X and Y are statistically independent. So for independent variables we have the result that the variance of the sum is the sum of the variances

$$\text{Var}\,(X+Y) = \sigma_X^2+\sigma_Y^2 \qquad 5.22$$

By the same argument it may be shown that

$$\text{Var}\,(aX+bY) = a^2\sigma_X^2+b^2\sigma_Y^2$$

The results of this section may be generalised to deal with several random variables X_1, X_2, X_3 ... by repeated application of 5.19, 5.20 and 5.22. If we form the sum

$$S = a_1X_1+a_2X_2+a_3X_3+ \dots$$

Signals and Information

then

$$\mu_s = a_1\mu_{X_1} + a_2\mu_{X_2} + a_3\mu_{X_3} + \dots \qquad 5.23$$

while, if $X_1, X_2, X_3 \dots$ are mutually independent

$$\sigma_s^2 = a_1^2\sigma_{X_1}^2 + a_2^2\sigma_{X_2}^2 + a_3^2\sigma_{X_2}^2 + \dots \qquad 5.24$$

5.10 COVARIANCE AND CORRELATION

When X and Y are not statistically independent the last term in euqation 5.21 does not vanish. The quantity

$$\text{Cov}\,(X,\,Y) = E[(X-\mu_X)(Y-\mu_Y)] = E(XY) - \mu_X\mu_Y$$

is thus a measure of their statistical dependence and is termed the *covariance* of X and Y. Closely related is the correlation coefficient, defined by

$$\varrho(X,\,Y) = \frac{\text{Cov}\,(X,\,Y)}{\sigma_X\sigma_Y}$$

Equation 5.22 now reads

$$\text{Var}\,(X+Y) = \sigma_X^2 + 2\varrho\sigma_X\sigma_Y + \sigma_Y^2$$

and it follows that

$$\text{Var}\left(\frac{X}{\sigma_X} \pm \frac{Y}{\sigma_Y}\right) = 1 \pm 2\varrho + 1 = 2(1 \pm \varrho)$$

but a variance cannot be negative, thus

$$|\varrho(X,\,Y)| \leqslant 1$$

It has been shown that, if two random variables are independent, their covariance vanishes, $\varrho = 0$, and they are said to be uncorrelated. The condition $\varrho = 0$ does *not* necessarily imply that the variables are statistically independent. There is, however, a particularly important case for which the converse *is* true—namely that in which X and Y both have Gaussian probability density functions (see Section 5.12 below). A proof that two Gaussian random variables which are uncorrelated are also statistically independent is beyond the scope of this book, but it is a useful property of Gaussian variables which will be required in the next chapter.

170

5.11 REPEATED INDEPENDENT TRIALS

In discussing probability we are always concerned with the outcomes of trials which, at least in principle, could be repeated indefinitely under identical conditions. Let us now suppose some sequence of trials has been made and that each outcome has yielded a value of some random variable, X. For example, a roulette wheel may be spun n times and the numbers produced written down, $x_1, x_2, x_3 \ldots$. Can anything be deduced about the sum of these numbers, or about their average? To tackle this problem we now view the complete series of n spins as one single trial and define a random variable M to be the arithmetical average of the numbers produced. Thus for one sequence

$$m = \frac{1}{n} \sum_{j=1}^{n} x_j$$

where each x_j is the value of the random variable X_j, defined as the number appearing at the jth spin of the trial.

We now make the assumption that the random variables X_j have identical probability functions with mean μ_X, variance σ_X^2 and also that they are statistically independent. Using equations 5.23 and 5.24 above, with $a_1 = a_2 = a_3 \ldots = 1/n$ we have

$$\mu_M = \mu_X \qquad \qquad 5.25$$

and

$$\sigma_M = \frac{1}{\sqrt{n}} \sigma_X \qquad \qquad 5.26$$

These results are of great importance in the theory of errors. When several observations have been made of the same physical quantity, the standard deviation (standard error) of their *average* diminishes as the square root of the number of observations.

We began this chapter by attaching numbers, which we called probabilities, to each possible outcome of a trial. The probabilities were subject to the conditions expressed by equations 5.1 and 5.2 and the theory has been developed by way of certain axiomatic rules for manipulating them. At this point we turn to the problem of determining probabilities experimentally. Let us focus our attention upon some event E which may occur with probability p and define a random variable I which takes the value 1 when E occurs and 0 otherwise. It then follows by definition (equation 5.9) that $\mu_I = p$. In this case $E(I^2) = p$ also (equation 5.10) so that 5.12 yields $\sigma_I^2 = p(1-p)$.

Suppose that a series of n independent trials is now made. The random variable M, defined above, then takes a value given by the number of times E occurred divided by the total number of trials, n. The value of M is thus the *relative frequency* of occurrence of E in that particular series and the mean of M is by equation 5.25, $\mu_M = p$. The probability that M will take a value differing by c from its mean value p is given by Chebyshev's inequality, namely

$$P(|M-p| > c) \leqslant \frac{\sigma_M^2}{c^2}$$

But by equation 5.26

$$\sigma_M^2 = \frac{\sigma_I^2}{n} = \frac{p(1-p)}{n}$$

and hence

$$P(|M-p| > c) \leqslant \frac{p(1-p)}{nc^2} \qquad\qquad 5.27$$

Now, no matter how small we make c, this probability may be made as small as we please by making n sufficiently large. Thus the probability that the relative frequency will have a value lying arbitrarily close to p approaches 1 when the trials are continued indefinitely. This is one form of the 'law of large numbers'. The event 'the relative frequency of occurrence of E lies in the range $p \pm c$' has a probability number which is almost 1 for large n. If we agree that events with a probability number equal to 1 are certain to occur, then we may estimate the probability of E by performing a long series of trials and noting the fraction of times E occurs. It is then almost certain that this fraction lies very close to p.

The theory of probability may be developed by starting with a definition of probability as the limiting value of the relative frequency of occurrence of an event in an indefinitely long sequence of trials. The account in this chapter follows the one most commonly adopted. Treatments, or references to treatments, of the difficult and controversial subject of the foundations of probability theory will be found in the suggestions for further reading listed at the end.

5.12. SOME SPECIAL PROBABILITY DISTRIBUTIONS

We conclude this chapter with a brief account of some particular probability distributions which will be important in later work. The first is the binomial distribution which concerns the probability of obtaining a certain number of 'successes' in a given number of trials whose outcomes are classified into the two possible events labelled 'success' and 'failure'. The binomial distribution is basic to the theory of information transfer over a binary channel. The Poisson distribution will be seen to have an application in the study of shot noise. The Gaussian or normal distribution describes a random variable which is itself the sum of a large number of independent variables and is thus appropriate to the study of thermal and other noise fluctuations which are the cumulative result of a huge number of atomic and electronic disturbances. Closely related to it is the Rayleigh distribution.

5.12.1. Binomial Distribution

Consider a series of n trials in each of which the probability of occurrence of some event E is p, successive trials being independent. What is the probability that E will occur r times? To answer this question let us view the whole series of trials as a single experiment. For the whole experiment we now define the events E_k such that E_k has occurred if E happened at the kth trial of the experiment. Then n random variables I_k may be defined such that $I_k = 1$ when E_k occurs and $I_k = 0$ otherwise. For example, in an experiment consisting of six trials one of the 2^6 possible outcomes might be recorded as

$$1\ 0\ 1\ 1\ 1\ 0$$

in which events E_1, E_3, E_4, and E_5 have occurred and E_2 and E_6 have not occurred. We assume that $P(E_k) = p$ and $P(\text{not } E_k) = 1 - p$ for all k, and that the events E_k are statistically independent. The product law for independent events may therefore be used to evaluate the probability of this particular outcome, namely

$$p(1-p)\ p\ p\ p(1-p) = p^4(1-p)^2$$

For any of the possible outcomes, the random variable $R = \sum_k I_k$ takes a value equal to the number of times E occurred in the experiment. The particular outcome considered above has probability $p^4(1-p)^2$ and $r = 4$, but the probability $P(R = 4)$, that R takes the value 4, is different, for there are 15 different ways of obtaining a sequence containing 4 'successes' and 2 'failures'. Hence, by the sum rule

$$P(R = 4) = 15p^4(1-p)^2$$

is the probability that E could occur four times in six trials.

In general, for a series of n trials, r events E may be obtained in

$$C_r^n = \frac{n!}{r!\,(n-r)!}$$

different ways, so that

$$P(r) = \frac{n!}{r!\,(n-r)!}\,p^r(1-p)^{n-r} \qquad 5.28$$

Repeated independent trials which involve 'success' (E occurs) or 'failure' (E does not occur) at each trial are often termed Bernoulli trials, and the probability function of equation 5.28 the Bernoulli distribution, after James Bernoulli (1654—1705). An alternative name is the 'binomial distribution' because $P(r)$ is also the $(r+1)$th term in the binomial expansion of $[p+(1-p)]^n$, from which it is also clear that

$$\sum_{r=0}^{n} P(r) = 1$$

as must be the case (equation 5.8).

The mean and variance of the binomial random variable may be found as follows. We note that each I_k has mean $\mu_{I_k} = p$ and variance Var$(I_k) = p(1-p)$. But R is defined as

$$R = \sum_k I_k$$

so that the results of section 5.9 yield

$$\mu_R = np \qquad 5.29$$

$$\sigma_R^2 = np(1-p) \qquad 5.30$$

Fig. 5.1 shows the binomial probability function $P(r)$ for $p = 0.7$ and $n = 20, 100,$ and 1000. The peak becomes relatively more and more narrow as n increases, illustrating the law of large numbers. The diagram also indicates the probability that r will lie in the range $n \times (0.7 \pm 0.05)$

5.12.2. Poisson Distribution

If in the binomial distribution $p \ll 1$ but n is sufficiently large for the mean, np, to be significant, the probability function $p(r)$ may be shown to take the approximate form

$$p(r) = \frac{\mu^r e^{-\mu}}{r!} \qquad 5.31$$

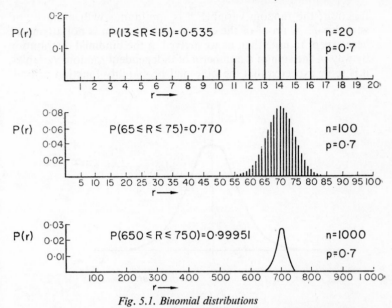

Fig. 5.1. Binomial distributions

where $\mu = np$ is the mean of the distribution. A distribution given exactly by equation 5.31 is called a Poisson distribution. The variance, from 5.30, becomes $\sigma^2 = np = \mu$ and this is thus equal to the mean. The Poisson distribution finds important applications in discussing events which are random in time, such as the counts of a Geiger counter, the number of calls arriving at a telephone switchboard or the number of electrons traversing a saturated diode in a given time interval.

5.12.3 Gaussian Distribution

Perhaps the most famous theorem in probability theory is the 'central limit theorem'. This asserts that under certain (but usually valid) conditions, the sum of a large number of independent random variables has a probability function approximating to the form shown in Fig. 5.2

$$p(x) = \frac{1}{\sqrt{(2\pi\sigma^2)}}\, e^{-(x-\mu)^2/2\sigma^2} \qquad 5.32$$

This, unlike the previous two distributions, is a probability density

175

function; the random variable X is continuous with mean μ and variance σ^2. A proof of the central limit theorem is not attempted here except to note that, as we arrived at the binomial distribution by way of the sum of a number n of independent random variables, it is to be expected that the binomial probability function will ap-

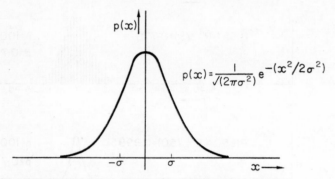

Fig. 5.2. The Gaussian or normal distribution ($\mu = 0$)

proximate to the normal form when n is made sufficiently large. Similarly if the mean, μ, of a Poisson distribution is made large then this probability function also becomes approximately Gaussian. The reader is invited to demonstrate these special cases for himself using Stirling's approximation for the factorial.

A particularly important property of Gaussian random variables is that the sum of two independent Gaussian variables is itself Gaussian. Suppose that X_1 and X_2 are two such variables with means μ_1 and μ_2 and variances σ_1^2 and σ_2^2. If we imagine X_1 and X_2 to be represented by rectangular coordinates (Fig. 5.3) then $p(x_1, x_2)$ $dx_1 dx_2$ represents the probability that X_1 and X_2 have values in the range represented by points within an area $dx_1 dx_2$ at the point (x_1, x_2). We now note that the sum $Z = X_1 + X_2$ has a constant value Z along the line $x_1 + x_2 = z$. The probability that Z will have a value between z and $z + \delta z$ is then given by

$$\int \int p(x_1, x_2) \, dx_1 \, dx_2$$

where the integration extends over the area between the lines AB and CD (Fig. 5.3). Putting in the limits and remembering that

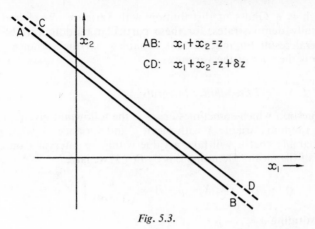

AB: $x_1 + x_2 = z$

CD: $x_1 + x_2 = z + \delta z$

Fig. 5.3.

$p(x_1, x_2) = p(x_1) p(x_2)$ since X_1 and X_2 are assumed independent

$$p(x) \, \delta z = \int_{-\infty}^{\infty} p_1(x_1) \left[\int_{z-x_1}^{z-x_1+\delta z} p_2(x_2) \, dx_2 \right] dx_1$$

in the limit the expression in square brackets becomes $p_2(z - x_1) \, dz$ and thus

$$p(z) = \int_{-\infty}^{\infty} p_1(x_1) \, p_2(z - x_1) \, dx_1$$

This is a general result for the probability density function of the random variable Z, defined as the sum of any two independent random variables X_1 and X_2 with probability density functions $p_1(x_1)$ and $p_2(x_2)$, respectively. The required function $p(z)$ is the convolution integral of $p_1(x)$ and $p_2(x)$. We have already noted (Chapter 2, Problem 12) that the convolution integral of two Gaussian functions is itself Gaussian, so that it is clear at this point that the sum of two independent Gaussian random variables is Gaussian. It is instructive to proceed in detail, however, for the case $\mu_1 = \mu_2 = 0$ (finite means correspond only to a shift of the origin in Fig. 5.3)

$$p(z) = \frac{1}{2\pi\sigma_1\sigma_2} \int_{-\infty}^{\infty} \exp -\frac{1}{2} \left[\frac{x_1^2}{\sigma_1^2} - \frac{(z - x_1)^2}{\sigma_2^2} \right] dx_1$$

$$= \frac{1}{\sqrt{[2\pi(\sigma_1^2 + \sigma_2^2)]}} \exp -\frac{z^2}{2(\sigma_1^2 + \sigma_2^2)}$$

177

which is a Gaussian distribution with variance $\sigma_1^2 + \sigma_2^2$ and incidentally demonstrates, for these particular random variables, the general result for independent variables that the variance of the sum is the sum of the variances.

5.12.4. Normal Probability Integrals

A question which sometimes arises is the following: given a Gaussian random variable X with mean μ and variance σ^2, what is the probability that it will take a value within an interval s on either side of the mean? Denoting this by $P(\pm s)$ we have

$$P(\pm s) = P(\mu - s \leqslant X \leqslant \mu + s) = \frac{1}{\sqrt{(2\pi\sigma^2)}} \int_{\mu-s}^{\mu+s} e^{-(x-\mu)^2/2\sigma^2}\, dx$$

Substituting $u = x - \mu$

$$P(\pm s) = \frac{1}{\sqrt{(2\pi\sigma^2)}} \int_{-s}^{s} e^{-u^2/2\sigma^2}\, du \qquad 5.33$$

and since the integrand is even

$$P(\pm s) = \sqrt{\left(\frac{2}{\pi\sigma^2}\right)} \int_{0}^{s} e^{-(u^2/2\sigma^2)}\, du$$

This important definite integral is commonly tabulated in one of two ways. The first sets $y = u/\sigma$ and tabulates values of the integral of a normal distribution with zero mean and unit variance, thus

$$P(\pm s) = \sqrt{\left(\frac{2}{\pi}\right)} \int_{0}^{s/\sigma} e^{-y^2/2}\, dy \qquad 5.34$$

Alternatively, one may set $t = u/\sigma\sqrt{2}$, whereupon

$$P(\pm s) = \frac{2}{\sqrt{\pi}} \int_{0}^{s/\sqrt{2}\sigma} e^{-t^2}\, dt$$

$$= \operatorname{erf}\left(\frac{s}{\sqrt{2}\sigma}\right) \qquad 5.35$$

where erf (η) is the so-called error function. For large η it is of interest to note that

$$\operatorname{erf}(\eta) \approx 1 - \frac{e^{-\eta^2}}{\sqrt{(\pi\eta^2)}} \qquad 5.36$$

5.12.5 *Rayleigh Distribution*

Consider the random variable R, defined as

$$R = \sqrt{(X^2 + Y^2)}$$

where X and Y are independent Gaussian random variables, each with zero mean and variance σ^2. The chance that a single trial will result in values of X and Y lying within x, $x + dx$ and y, $y + dy$, respectively, is given by

$$p(x, y)\, dx\, dy = \frac{1}{\sqrt{(2\pi\sigma^2)}}\, e^{-x^2/2\sigma^2} \times \frac{1}{\sqrt{(2\pi\sigma^2)}}\, e^{-y^2/2\sigma^2}\, dx\, dy$$

$$= \frac{1}{2\pi\sigma^2}\, e^{-(x^2 + y^2)/2\sigma^2}\, dx\, dy$$

$$= \frac{1}{2\pi\sigma^2}\, e^{-r^2/2\sigma^2}\, dx\, dy.$$

Integrating within the annular region r, $r + dr$ shown in Fig. 5.4 yields

$$p(r)\, dr = \frac{r}{\sigma^2}\, e^{-r^2/2\sigma^2}\, dr \qquad\qquad 5.37$$

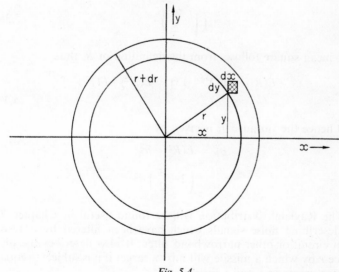

Fig. 5.4.

179

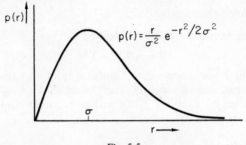

Fig. 5.5.

This is the Rayleigh distribution and has the form sketched in Fig. 5.5.

The mean of the distribution may be found with the aid of the definite integral

$$\int_0^\infty x^2 e^{-a^2 x^2}\, dx = \frac{\sqrt{\pi}}{4a^3}$$

to give

$$\mu_R = \int_0^\infty \frac{r^2}{\sigma^2} e^{-r^2/2\sigma^2}\, dr$$

$$= \left(\sqrt{\frac{\pi \sigma^2}{2}} \right) \qquad 5.38$$

The mean square follows from the definition of R, thus

$$E(R^2) = E(X^2 + Y^2) = E(X^2) + E(Y^2)$$

$$= 2\sigma^2 \qquad 5.39$$

and hence the variance σ_R^2 is given by

$$\sigma_R^2 = E(R^2) - \mu_R^2$$

$$= \left(2 - \frac{\pi}{2} \right) \sigma^2 \qquad 5.40$$

The Rayleigh distribution will be found useful in Chapter 7, in describing noise signals which have been filtered by a resonant circuit or other narrowband filter. It also describes the distance r by which a missile will miss a target if it is subject to equal aiming errors in x and y directions.

Further Reading

FELLER, W., *Introduction to Probability Theory and Application*, vol. 1 (Wiley, New York, 1957)

BRUNK, H. D., *Introduction to Mathematical Statistics*, (Ginn, Boston, 1960)

FREEMAN, H., *Introduction to Statistical Inference*, (Addison-Wesley, New York, 1963)

KEEPING, E. S., *Introduction to Statistical Inference*, (Van Nostrand, Princeton, 1962)

A lively account of the relative frequency definition of probability and the law of large numbers is given by R. von Mises in *Probability, Statistics and Truth* (English translation) (Hodge, London, 1939)

Many of the texts devoted to statistical communication theory contain tutorial material on probability theory; see, for example:

WOODWARD, P. M., *Probability and Information Theory, with Applications to Radar* (Pergamon, London, 1953)

DAVENPORT, W., and ROOT, W., *Introduction to the Theory of Random Signals and Noise* (McGraw-Hill, New York, 1958)

PROBLEMS

1. A complex electronic system is checked by means of a certain test. From experience it is known that there is a 99·9% chance that the system is operating correctly. On a certain day, because of a fault in the test equipment, it happens that if the system is working incorrectly there is a 1% chance that the test will indicate no fault, while if the system is working correctly there is a 0·5% chance that the test will indicate a fault. The test indicates a fault. Show that the probability that the fault is real is approximately $\frac{1}{6}$.

2.

	A_1	A_2	A_3	A_4
B_1	$\frac{1}{8}$	$\frac{1}{4}$	$\frac{1}{16}$	$\frac{1}{16}$
B_2	$\frac{3}{32}$	$\frac{3}{16}$	$\frac{7}{128}$	$\frac{5}{128}$
B_3	$\frac{1}{32}$	$\frac{1}{16}$	$\frac{1}{128}$	$\frac{3}{128}$

The outcomes of a certain trial are classified into mutually exclusive events A and also into mutually exclusive events B. The joint probabilities $P(A_j, B_k)$ are as shown in the table. Calculate the marginal probabilities $P(A_j)$, $P(B_k)$. Which pairs of events are independent and which are not? Calculate $P(A_2 | B_1)$, $P(A_2 | B_2)$, $P(A_2 | B_3)$ and $P(B_1 | A_2)$ by inspection of the table and the last one also from Bayes's formulae.

2. (*cont.*) Two random numbers X and Y are defined such that X takes the value 1 if A_1 or A_4 occur, 0 if A_2 occurs and 2 if A_3 occurs while Y takes the value 0 if B_2 occurs and 1 otherwise. Find μ_X, μ_Y, $E(X+Y)$, $E(XY)$, Cov (XY). Are X and Y independent?

3. A continuous random variable V may take any value with equal probability in the range 0 to Q. Find μ_v, $E(V^2)$ and σ_v^2.

4. Noise on a teleprinter link results in symbols being mistaken at the receiver. It is decided to limit messages to 80 symbols and to precede these with 20 test symbols on transmission. If the test symbols are received correctly the message is accepted for decoding. What is the probability that if two errors exist in the 100 symbols, the message will be accepted?

5. A measurement of a certain noise voltage yields a Gaussian random variable with zero mean and variance $10^{-11}\ V^2$. What is the probability that a sample measurement exceeds $5\ \mu V$?

6. Show that $\displaystyle\int_{-\infty}^{\infty} (x-a)^2\, p(x)\, dx$ is a minimum when $a = E(x)$.

7. A probability distribution function $p(x)$ has Fourier transform $P(\omega)$. Verify that $E(X) = j\,P'(0)$ and Var $(X) = [P'(0)]^2 - P''(0)$. Use these relations to evaluate the mean and variance of a Gaussian distribution.

Chapter 6

Noise

6.1 INTRODUCTION

A tremendous variety of natural phenomena contributes to the
e.m.f. induced in a radio aerial. It is estimated that, over the whole
of the Earth's surface, approximately one hundred lightning strokes
occur per second, each with an energy spectrum extending up to
10 MHz or more. The Sun is by no means a quiet object and random noise signals from this and other cosmic sources will be added
to the received signal. At frequencies above about 1000 MHz noise
of thermal origin from the Earth's surface and the atmosphere can
play a significant role.

There is also man-made noise from electrical machinery or ignition systems, or due to interference from signals occupying an adjacent channel in the spectrum, or due to mains hum which, together
with impulses from electrically charged dust or sand particles
striking the aerial, might in principle be eliminated. The overall
effect is to add to the desired message signal a random component
which is generally termed *additive noise*. The received carrier may
also exhibit fluctuations in intensity, generally due to the arrival
of two or more signal components by different modes of propagation which superpose with randomly varying phase difference. The
term *multiplicative disturbances* is appropriate for these effects; they
complicate the received signal still further, but will not be discussed
here.

Because of attenuation, any communication link, whether by

cable or radio, requires the use of amplifiers at the receiver and these themselves add more noise to the signal. How much noise is added from all sources, and the extent to which it can be combatted by the design of the system, will determine the transmitter power needed or, if this is fixed, the distance over which reliable communications can be established. It is therefore essential to have some understanding of the basic features of noise sources and noise signals, which are the common enemy of both the communications engineer and scientific workers in other fields who are concerned with the detection of weak signals.

This chapter is devoted to an elementary review of some of the sources of additive noise and how they combine. Its purpose is to provide a background for the analysis of systems which handle noisy signals, to be given in the next chapter. The design of low-noise equipment is an important and specialised study, but it would lead us away from the main theme of this text.

6.2 THERMAL NOISE IN RESISTORS. ADDITION OF NOISE POWERS

Anyone who has turned up the gain control of a sensitive amplifier knows something of electrical noise. If the output is connected to a loudspeaker a steady hiss is heard, while an oscilloscope displays a random waveform which cannot be synchronised to show

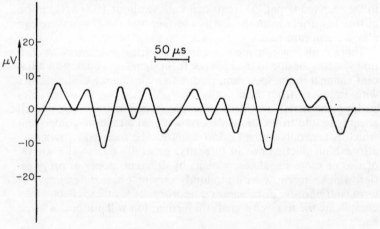

Fig. 6.1(a).

any repetitive pattern. The noise output can be significantly reduced by short circuiting the input terminals of the amplifier, but if a resistor is connected across the input the noise increases again, becoming larger the larger the resistance used. This shows that a resistor is a source of a randomly fluctuating e.m.f. The phenomenon was first investigated experimentally by J. B. Johnson[1] and theoretically by H. Nyquist[2] in 1928. The physical origin of the noise lies in the thermal agitation of the charge carriers in the resistive material, as may be demonstrated by cooling the resistor, whereupon the noise fluctuations are reduced. Noise of this type is referred to either as *thermal noise* or as *Johnson noise*.

Fig. 6.1 shows waveforms traced from single-sweep oscillographs obtained by connecting a $100 \text{ k}\Omega$ resistor across the input terminals of an amplifier having low internal noise and a large input impedance. Trace (a) obtained with a 20 kHz low-pass filter inserted following the amplifier. Trace (b) was obtained under the same conditions but with the time base running five times more slowly. In trace (c) the time scale is the same as in (a) but the 20 kHz filter was replaced with one having a cut-off frequency five times higher, at 100 kHz. There is a similarity between (b) and (c) although the wider bandwidth has resulted in a larger r.m.s. level. The final trace (d), of which more will be said in the next chapter, shows the effect of inserting a band-pass filter having a pass band from 17·5 to 20 kHz.

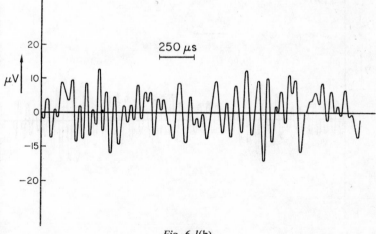

Fig. 6.1(b).

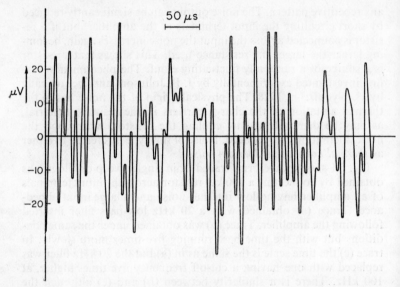

Fig. 6.1(c).

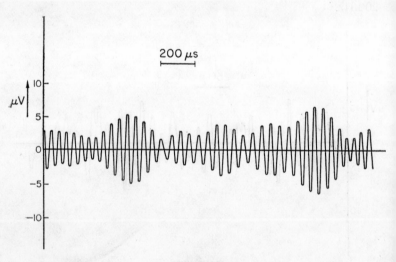

Fig. 6.1(d).

The mean square level of such signals may be measured by connecting the amplified output to an instrument such as a thermal milliammeter which effectively measures the quantity

$$N = \frac{1}{T} \int_{-T/2}^{T/2} v^2(t) \, dt$$

and if the response time of the instrument (and hence T) is sufficiently long, a sensibly steady value is recorded which is proportional to the mean noise power. Careful experiments show that the mean square voltage is linearly proportional to the value of the resistance, the temperature of the resistor and the bandwidth of the measuring system (strictly the 'noise bandwidth' defined in Section 6.6).

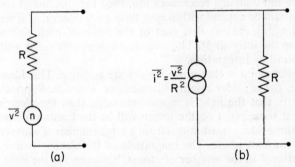

Fig. 6.2. Thevenin and Norton representations for Johnson noise

To take account of thermal noise in circuit calculations, a randomly fluctuating e.m.f. of mean square value $\overline{v^2}$ may be included in series with each resistor, as in Fig. 6.2(a). Alternatively, the Norton equivalent of Fig. 6.2(b) may be used. The noise voltage or current appearing at some point in a resistive network will, at any particular instant, be the linear superposition of those produced by the separate noise e.m.f.'s according to the usual laws of circuit analysis. The magnitudes of the noise voltages, however, are not known at any given instant so that this kind of calculation, although valid in principle, cannot be performed.

We may, however, regard the noise source e.m.f.'s at some instant as independent random variables and the linear superposition of independent random variables has been treated in the previous chapter. There can be no doubt that the separate noise e.m.f.'s in different resistors are independent. The appropriate equations are

187

5.23 and 5.24. Since noise quantities have zero means, the variance is identical to the mean square so that

$$\overline{v^2} = a_1^2 \overline{v_1^2} + a_2^2 \overline{v_2^2} + a_3^2 \overline{v_3^2} + \ \ldots \qquad 6.1$$

where v_1, v_2, v_3 are the several noise e.m.f.'s in the circuit and the coefficients a_1, a_2, a_3, etc. are those calculated by normal linear circuit theory.

Other sources of noise, such as those generated in active devices or from external sources, may be included in the same way. In this way the *mean square noise voltage* at some point in a circuit may be calculated, provided the mean square values of the individual sources (thermal or otherwise) are known. If the circuit is not purely resistive but contains reactances too, then the method of equation 6.1 may still be applied within each frequency interval δf if we interpret v_1^2, v_2^2, v_3^2, etc., as that part of the noise of each source contained in the interval δf. The overall mean square noise voltage is then found by integration.

The key point is that noise *powers* are additive. This idea is not strange, for consider a lamp illuminating a screen. Supposing for simplicity that the light is monochromatic, then the electric field vector at some point on the screen will be the linear superposition of the sinusoidal contributions from a large number of atoms radiating in random phase. The magnitude of the electric vector is not proportional to the number of atoms, however, for the light intensity is given by the mean square value of the electric field and it is the intensity which is proportional to the number of atoms. Two lamps give twice the light flux and not four times the light flux of one lamp. The reason lies in the fact that the light waves from individual atoms have phases which are statistically unrelated. If atoms can be induced to emit at identical frequency *and* phase, as in the laser, the result is startlingly different.

The significant quantity for thermal noise, therefore, is the $\overline{v^2}$ of Fig. 6.2 and thermodynamic arguments may be invoked to determine how this depends on the resistance and temperature. The following partial analysis illustrates the ideas of the preceding four paragraphs.

Consider two resistors R_1 and R_2 connected together as shown in Fig. 6.3(a). Each resistor is delivering power to the other by virtue of the noise e.m.f.'s $\overline{v_1^2}$ and $\overline{v_2^2}$. If the resistors are at the same temperature, then in a given time interval each must, on average, receive as much energy as it delivers to its partner, otherwise the second

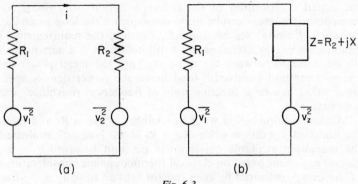

Fig. 6.3.

law of thermodynamics would be violated. Suppose that at some instant the noise e.m.f.'s are v_1 and v_2, then the instantaneous current is

$$i = \frac{v_1}{R_1+R_2} - \frac{v_2}{R_1+R_2}$$

and hence from the argument above

$$\overline{i^2} = \frac{\overline{v_1^2}}{(R_1+R_2)^2} + \frac{\overline{v_2^2}}{(R_1+R_2)^2}$$

the mean power dissipated in R_2 is

$$\overline{i^2}R_2 = \frac{\overline{v_1^2}R_2}{(R_1+R_2)^2} + \frac{\overline{v_2^2}R_2}{(R_1+R_2)^2} \qquad 6.2$$

The first term in this equation represents the mean power received by R_2 from R_1. A similar calculation yields $\overline{v_2^2}R_1/(R_1+R_2)^2$ for the mean power received by R_1 from R_2. Since these must be equal

$$\frac{\overline{v_1^2}R_2}{(R_1+R_2)^2} = \frac{\overline{v_2^2}R_1}{(R_1+R_2)^2}$$

or

$$\frac{\overline{v_1^2}}{R_1} = \frac{\overline{v_2^2}}{R_2}$$

and hence we conclude that for thermal noise

$$\overline{v^2} \propto R \qquad 6.3$$

189

the actual composition of the resistor—carbon, wirewound, or semiconductor, etc.—can be of no consequence (so long as no d.c. current is flowing: see Section 6.5). Further, the requirement of equilibrium energy exchange must still hold even if a narrowband filter is inserted between the resistors. Thus the mean power exchanged per unit bandwidth (and hence the power density spectrum of v) can be a function only of frequency, resistance, and temperature.

Maximum noise power will be exchanged between R_1 and R_2 in the matched condition when $R_1 = R_2 = R$. Nyquist[2] evaluated the maximum available noise power per unit bandwidth by an elegant argument based on classical thermodynamic considerations of the energy delivered by each resistor into an initially noise-free transmission line joining them. The argument will not be reproduced here and the reader is urged to read the original paper for himself. The result is that a resistor will deliver a power kTB watts within a bandwidth B Hz into a matched load. From equation 6.2, this power is $\overline{v^2}R/(R+R)^2 = \overline{v^2}/4R$ and hence

$$\overline{v^2} = 4kTRB \qquad\qquad 6.4$$

where $k = 1\cdot38.10^{-23}$ J K^{-1} is Boltzmann's constant.

This expression is independent of frequency, that is to say the power density spectrum is uniform (but see below). Noise like this is said to be 'white'. Putting in some figures, if $T = 292°$K, $R = 10$ kΩ, and $B = 1$ kHz, $\overline{v^2} = 1\cdot6.10^{-13}$ V^2 corresponding to an r.m.s. fluctuation of $0\cdot4$ μV.

It is first alarming that, according to equation 6.4, the noise power increases without limit as $B \rightarrow \infty$ (cf. the 'ultra-violet catastrophe' in classical theories of black body radiation). As Nyquist pointed out, however, when quantum statistics are used, the expression kTB should be replaced by

$$\int_{f_1}^{f_1+B} \frac{hf\,\mathrm{d}f}{\mathrm{e}^{hf/kT} - 1}$$

where $hf/\mathrm{e}^{hf/kT}-1) \approx kT$ if $f \ll kT/h$. At 20°C, $kT/h = 6\cdot12.10^{12}$ Hz, so equation 6.4 may safely be used in practical cases at normal temperatures up to frequencies at least as high as those at which resistors behave like lumped circuit elements.

Now suppose we replace the resistor R_2 as in Fig. 6.3(b), by an impedance Z, which at some frequency f may be written $Z = R_2 + jX$.

Equating, as before, the power transferred each way

$$\frac{R_2 \,\overline{\delta v_1^2}}{(R_1+R_2)^2+X^2} = \frac{R_1 \,\overline{\delta v_2^2}}{(R_1+R_2)^2+X^2}$$

In this equation $\overline{\delta v_1^2}$ is obtained from equation 6.4 with $B = \delta f$ a narrow frequency interval centred on f. The left-hand side is the power delivered into Z from R_1 within the frequency interval δf. (We recall that a pure reactance cannot absorb power.) The right-hand side is the power delivered to R_1, assuming a noise source $\overline{\delta v_z^2}$ for Z. Setting $\overline{\delta v_1^2} = 4kTR \,\delta f$ we have

$$\overline{\delta v_2^2} = 4kTR_2 \,\delta f = \overline{\delta v_z^2}$$

which shows that the noise associated with Z is due only to the resistive part of the impedance. It follows that a lossless capacitor or inductor is not a source of noise fluctuations. Thermal noise sources are always power absorbers as well.

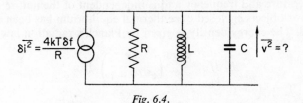

Fig. 6.4.

This section is concluded with an example which illustrates the way noise sources enter into circuit calculations. The problem, illustrated in Fig. 6.4, is to find the mean square voltage $\overline{v^2}$, across the parallel resonant circuit L, C, R. The contribution to $\overline{v^2}$ in a frequency interval δf at f will be

$$Z^2 \,\overline{\delta i^2} = \frac{4kT \,\delta f}{R} \cdot \frac{R^2}{1+Q^2(f/f_0-f_0/f)^2}$$

where $Q = 2\pi f_0 CR$ and f_0 is the resonant frequency. Adding mean square contributions at all frequencies

$$\overline{v^2} = 4kTR \int_0^\infty \frac{\mathrm{d}f}{1+Q^2(f/f_0-f_0/f)^2} \qquad 6.5$$

191

The integral can be evaluated by the technique of contour integration (see also Pierce[3]). The result is $\pi f_0/2Q$, hence

$$\overline{v^2} = 2\pi f_0 kT/Q = kT/C \qquad 6.6$$

It follows that the mean energy stored in C is $C\overline{v^2}/2 = kT/2$; the mean energy stored in L will be the same. Thus the total energy stored in the LC circuit is kT, which is the expected classical result for a simple harmonic oscillator at temperature T.

6.3 NOISE RECEIVED BY AN AERIAL. SKY NOISE TEMPERATURE

Consider an aerial, matched to a resistive load, and contained within a constant-temperature enclosure. The familiar thermodynamic arguments relating to the energy balance between objects in such an enclosure may then be applied. In particular, the radiation energy per unit volume within a frequency interval df depends only on temperature and frequency and is independent of the nature of the walls or objects enclosed once thermal equilibrium has been established. The energy density is given by Planck's radiation law

$$E_f \, df = \frac{8\pi f^2}{c^3} \left(\frac{hf}{e^{hf/kT} - 1} \right) df$$

$$\approx \frac{8\pi f^2 kT}{c^3} \, df \qquad f \ll \frac{kT}{h}$$

Also the power radiated and absorbed per unit area by the walls of the cavity may be calculated and this is identical to that which would be radiated by a perfectly absorbing, or 'black', body at the same temperature, whether inside the enclosure or not. This power is given by

$$dP = \frac{2\pi h f^3}{c^2(e^{hf/kT} - 1)} \, df \approx \frac{2\pi f^2 kT}{c^2} \, df \qquad \text{watts per unit area}$$

If we couple the aerial by way of a transmission line to its load resistor and place this resistor in a second enclosure at the same temperature, then it is clear from the second law of thermodynamics that the resistor supplies a mean power to the aerial equal to that received by the aerial from the radiation field in its enclosure and

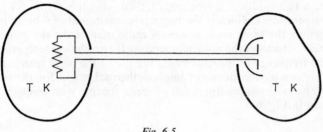

Fig. 6.5.

transmitted to the resistor. This is illustrated in Fig. 6.5. The power supplied by the load resistor within a bandwidth B has already been stated to be kTB (provided the temperature is not so low as to make the Planck correction significant). It follows that the power received by the aerial is

$$P = kTB \qquad\qquad 6.7$$

(For an alternative and more direct derivation see, for example, Jolly.[4])

The aerial is receiving noise power from that part of the enclosure intercepted by its beam width. If it were removed from its enclosure and directed towards a black radiator at temperature T, the noise power would still be given by equation 6.7.

The first observations of extraterrrestrial radio waves were made by K. G. Jansky in 1932. Jansky was investigating the background noise at around 20 MHz which was hampering communications on the 15 m band. His observations, using a directional aerial, clearly indicated a component of extraterrestrial origin which was observed to be most intense from the direction of the centre of the galaxy. It was thus a communication engineering problem which gave rise to the whole new science of radio astronomy.

The sky does not, of course, radiate like a black body. There are relatively intense localised radio sources which are observed against a background whose spectrum is continuous apart from a single line at 21 cm (1420 MHz) arising from a hyperfine transition in atomic hydrogen. The continuous galactic emission rises slowly in intensity as the direction of the galactic plane is approached and passes through a sharp maximum in a band occupying a few degrees on either side of the galactic equator. Equation 6.7 provides a convenient measure for the noise signals. If a power P is received by a (highly directional) aerial matched to a receiver of relatively narrow bandwidth B at frequency f, then substitution into 6.7 gives a temperature T which is the temperature required of a black body to deliver the same noise power. A radio map of the sky generally shows contours of the noise temperature (or brightness temperature) at the frequency of measurement. The use of the word temperature in this connection does not imply a thermal origin for the noise. The observed temperatures fall off quite sharply with frequency as shown in Fig. 6.6.

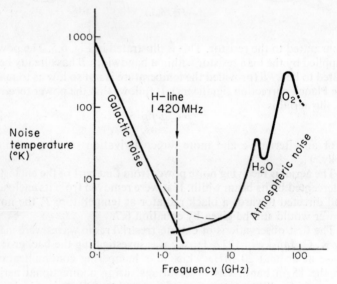

Fig. 6.6. Sky noise temperature (After Jolly[1])

In conclusion we note that, as thermal noise is always associated with the absorption of power, a lossy medium in the path of a radio wave provides a source of additional noise, depending on the temperature of the absorber. Atmospheric absorption, due primarily to water vapour and molecular oxygen, becomes appreciable at frequencies in excess of about 10^9 Hz and the noise temperature of the sky rises again from a minimum of a few degrees Kelvin at about 3.10^9 Hz. Losses in aerial feeders and couplings will also add to the received noise power.

6.4 SHOT NOISE

When the anode potential of a thermionic diode is made sufficiently large, the current flowing saturates to a value determined by the cathode temperature and work function. An electron having sufficient energy to overcome the surface potential barrier will cross to the anode producing a current pulse in the external circuit. It can be shown that for the case of a plane parallel diode (Fig. 6.7(a)) the current pulse is triangular and lasts for the transit time which is of order 10^{-10} s (Fig. 6.7(b)). The anode current thus consists of the superposition of many such pulses which are disposed randomly in time. The net current consequently fluctuates randomly about the mean value (Fig. 6.8) and these fluctuations are termed 'shot noise'.

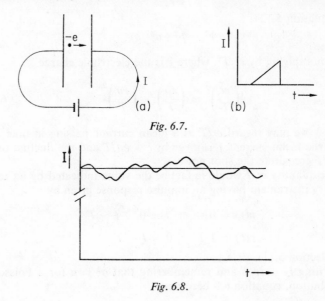

Fig. 6.7.

Fig. 6.8.

Shot noise is also generated in p–n junctions and transistors, in which conduction also occurs by electrons or holes in the tail of the energy distribution, which surmount a potential barrier. The phenomenon is particularly simple for the temperature-limited thermionic diode, however.

Suppose the chance that an electron will be emitted from the cathode in a time interval δt is $a\delta t$. If δt is sufficiently short, then

195

the probability that two electrons will leave the cathode in δt is negligible. The number of electrons emitted in some finite time T may then be regarded as the number of successes in $T/\delta t$ trials with a probability $a\delta t$ of success at each trial. The number of electrons emitted in time T will thus be a random variable of mean $\mu = aT$. Its probability function will be Poisson, since this is appropriate for the number of successes in a very large number of trials with only a small chance of success at each trial. Both of these assumptions become true as $\delta t \to 0$. The Poisson distribution, discussed in the previous chapter (equation 5.31), gives the probability that r electrons will be emitted in time T as

$$P(r) = \frac{e^{-\mu}\mu^r}{r!} \qquad 5.31$$

By equation 5.12

$$\overline{r^2} = \mu^2 + \sigma^2$$

and multiplying by e^2/T^2, where e is the electronic charge

$$\overline{\left(\frac{er}{T}\right)^2} = \left(\frac{e\mu}{T}\right)^2 + \left(\frac{e\sigma}{T}\right)^2 \qquad 6.8$$

Now we may regard er/T as a mean current passing in time T. The true mean current, i, is given by $i = e\mu/T$ and the fluctuations in er/T constitute the shot noise.

The quantity er/T would in fact be the value indicated by an imaginary instrument having an impulse response given by

$$h(t) = 0 \qquad t < 0, \quad t > T$$

$$h(t) = 1/T \qquad 0 < t < T$$

(see Section 3.11).

Writing $i_T = er/T$ and remembering that $\sigma^2 = \mu$ for a Poisson distribution, equation 6.8 becomes

$$\overline{i_T^2} = i^2 + \frac{ie}{T}$$

Referring to Fig. 6.8, we may interpret the left-hand side of this equation as the mean power delivered into a $1\,\Omega$ resistor by the current i_T, while the first term on the right represents the contribution due to the mean component and the second term is the shot

196

noise power

$$\overline{i_s^2} = \frac{ie}{T}$$

The quantity of primary interest, however, is not the noise power passed by the hypothetical 'averaging filter' but the power density spectrum of the noise component. It is plausible to assume that the power density is uniform up to a frequency of the order of the reciprocal of the transit time which we shall assume to be much less than T. The power density p_s of the noise component may then be taken as constant, so that the noise power passed by the filter is given by

$$\overline{i_s^2} = \frac{1}{2\pi} \int_{-\infty}^{\infty} p_s \, |H(\omega)|^2 \, d\omega$$

$$= \frac{1}{2\pi} \int_{-\infty}^{\infty} p_s \, \text{sinc}^2 \, (\omega T/2) \, d\omega$$

$$= \frac{p_s}{T}$$

where p_s is the two-sided power density, measured in $A^2 \, Hz^{-1}$ (see Section 2.18). Equating this expression for $\overline{i_s^2}$ with the value ie/T which was obtained from the statistical argument above, we have

$$p_s = ei$$

and thus the noise power received in an observation bandwidth B Hz is (equation 2.60)

$$\overline{i_s^2} = 2eiB \qquad\qquad 6.9$$

This is the usual expression for shot noise. Note that this expression is independent of the shape of the individual current pulses, lasting for time τ, provided the frequency at which observations are made is much less than $1/\tau$, so the result is not restricted to plane parallel diodes. If a resistor R is placed in series with a temperature-limited diode then the voltage fluctuations across R will be

$$\overline{v^2} = 2eiR^2B$$

For $i = 1$ mA, $R = 10$ kΩ and $B = 1$ kHz, $\overline{v^2} = 3 \cdot 2.10^{-11} \, v^2$, giving an r.m.s. fluctuation of about $5 \cdot 7 \, \mu V$.

Although the shot noise from a temperature-limited diode will not usually be part of the noise appearing in practical circuits, the simplicity of equation 6.9 means that this device can be used as a convenient and controllable source of white noise of accurately known power density for measurement purposes. Conversely, accurate measurements of shot noise have been used to measure the electronic charge. The noise power calculated from equation 6.9 will be reasonably accurate for frequencies up to several hundred MHz.

The discussion has been restricted so far to the temperature-limited region. If the diode current is space-charge limited, calculations become much more difficult. It is clear, however, that space charge will have the effect of reducing the current fluctuations. For suppose that the current emitted from the cathode is momentarily larger than the mean. This will have the effect of increasing the space charge and depressing the potential minimum in front of the cathode, thereby reducing the emission again. The space charge smooths out the fluctuations, successive electron transits being no longer statistically independent. Thomson, North and Harris[5] have shown that the shot noise is then given approximately by

$$\overline{i_s^2} = \left[\frac{9(1 - \pi/4)kT_c}{eV_a} \right] \times 2eiB \qquad 6.10$$

where T_c is the cathode temperature and V_a is the anode potential. The factor in square brackets is typically of the order 10^{-2} so the shot noise is significantly reduced.

In terms of the a.c. conductance of the diode, which is given by Child's law as $g = 3i/2V_a$, the last equation may be rewritten

$$\overline{i_s^2} = 2 \cdot 57 \, kT_c gB \qquad 6.11$$

The voltage fluctuations appearing across an anode load resistor R_L (Fig. 6.9(a)), may be calculated from the equivalent circuit of Fig. 6.9(b). In this the thermal noise of R_L has been included as well. The a.c. conductance of the diode cannot be neglected as it was in the temperature-limited case (see Problem 4).

A negative grid triode will also exhibit shot noise fluctuations in its anode current. An analysis is possible in terms of the equivalent diode description of triode operation. We quote only the result, which is an approximation for low-temperature (oxide-coated)

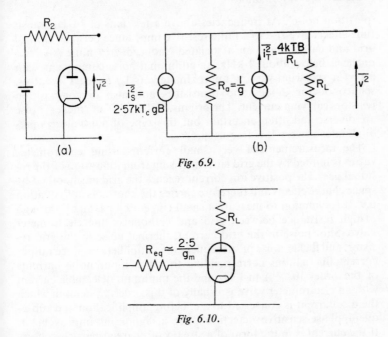

Fig. 6.9.

Fig. 6.10.

cathodes, that the shot noise may be represented as though produced by thermal noise in an equivalent resistor

$$R_{eq} \approx 2 \cdot 5 / g_m$$

in the grid circuit. This resistor (Fig. 6.10) is fictitious and does not load input circuits but adds noise $\overline{v^2} = 4kTR_{eq}B$ to any other noise voltages appearing between grid and cathode. Thermal noise in a resistive anode load, and from subsequent stages, may usually be neglected in comparison with the amplified noise from R_{eq} and the input grid resistor.

6.5 SOME OTHER SOURCES OF NOISE

The variety of noise sources existing in active circuits is too wide for any detailed discussion to be embarked on here. Besides thermal noise and shot noise, the random partition of the electron current between the positive grids and the anode of a multi-electrode valve gives rise to an additional component of anode current noise called

199

'partition noise'. At frequencies above a few tens of MHz, transit time effects give rise to a finite resistive impedance between control grid and cathode and an associated 'induced grid noise'. At frequencies below about 1 kHz, an additional noise component with a power spectrum varying approximately as $1/f$ appears in valves, semiconductor diodes and transistors and also in composition resistors carrying current. The origins of this '$1/f$' or 'flicker' noise are diverse and often uncertain but this type of noise always predominates at very low frequencies.

The measurement of very small currents using electrometer valves is limited by the grid current arising from ionisation of the residual gas. The positive ion current reaches the grid unsmoothed by space charge effects and therefore carries the shot noise of equation 6.9. It is common to measure a small current by passing it through a high resistance between grid and cathode of the electrometer valve. Shot noise in the grid current, thermal noise in the grid resistor, and flicker noise in subsequent d.c. amplifiers produce output current fluctuations corresponding to r.m.s. input noise currents of the order 10^{-14} A in 1 Hz and the minimum detectable current with an electrometer valve is usually of this order. Systems in which the d.c. current is converted to a.c. before amplification and subsequent phase-sensitive detection offer a significant improvement. If the current is in the form of a stream of charged particles, however, a much better performance is offered by secondary-emisson multipliers.

In a reverse-biased p–n junction, minority carriers from either side diffuse across the junction, giving a current pulse and then continue, becoming majority carriers. This gives the reverse saturation current the shot noise of equation 6.9. When the junction is forward biased additional noise components arise from the majority carriers, some of which are injected into the opposite region, giving a current pulse, and diffuse on as minority carriers which eventually recombine. Others cross the junction and diffuse back again, giving a doublet current pulse. At low frequencies the doublet pulse contributes little to the noise spectrum (cf. Transform C, Fig. 2.14) and the current carries noise which again has approximately the shot noise value (ignoring flicker noise). When forward biased, the noise is always less than, and at zero current is equal to, the thermal noise of the small-signal a.c. resistance.

Noise in a junction transistor can be described using an equivalent circuit involving three uncorrelated noise current generators connected base–emitter, emitter–collector, and collector–base.

A theoretical treatment is difficult; details are available in the more specialised texts, some of which are listed at the end of the chapter. The more recently developed field effect transistors show a much better noise performance although only relatively small gain–bandwidth products are possible. The noise generated in the gate–source junction due to leakage effects is small and noise figures (see below) of 1 dB or less are possible.

6.6 NOISE FIGURE, NOISE INPUT TEMPERATURE, NOISE BANDWIDTH

From the foregoing it is clear that some concise method is needed to describe the total additive noise introduced by an amplifier. This is usually done by comparing the signal-to-noise power ratios at the input and output. Let S_i be the signal power delivered into the input terminals, accompanied by noise power N_i. Let S_o and N_o be the signal and noise powers dissipated in the load at the output of the amplifier. The additional noise introduced by the amplifier means that the signal-to-noise power ratio at the input exceeds that at the output and accordingly a figure F may be defined as

$$F = \frac{(S_i/N_i)}{(S_o/N_o)} \qquad 6.12$$

If these observations are made in a sufficiently narrow bandwidth then F is called the *spot noise figure* at the frequency concerned. The spot noise figure will generally be a function of the signal source resistance as well as frequency.

A rather different description may be derived from equation 6.12 as follows. If the power gain of the amplifier is A then $S_i/S_o = 1/A$ and we have

$$F = N_o/AN_i \qquad 6.13$$

If we now write

$$N_o = A(N_i + N_A)$$

where N_A is interpreted as the noise power introduced by the amplifier referred to the input terminals, then

$$F = \frac{A(N_i + N_A)}{AN_i} = 1 + \frac{N_A}{N_i}$$

i.e., $N_A = N_i(F-1)$.

Signals and Information

It is customary to standardise N_i as the thermal noise from a matched source at 290°K, thus

$$N_i = 290 . k . \delta f$$

If we also write

$$N_A = T_A . k . \delta f$$

then T_A is called the *noise input temperature* which is related to the noise figure by

$$T_A = 290(F-1)°K \qquad 6.14$$

Over a wide bandwidth an average noise figure may be obtained using equation 6.13 which may be interpreted

$$\bar{F} = \frac{\text{noise output from amplifier}}{\text{noise expected from a noiseless amplifier of the same gain}}$$

For white input noise

$$dN_i = \text{const.} \times df$$

hence, using equation 6.13

$$\bar{F} = \frac{\displaystyle\int_0^\infty F . A . df}{\displaystyle\int_0^\infty A \, df} \qquad 6.15$$

An amplifier usually has a maximum gain somewhere near the centre of the pass band. If this power gain is A_0 then we may define a noise bandwidth such that a fictitious amplifier with uniform gain A_0 within the noise bandwidth and zero gain elsewhere would give the same output noise power as a noiseless amplifier having the same frequency characteristics as the real one. Thus

$$B_N = \frac{1}{A_0} \int_0^\infty A \, df \qquad 6.16$$

For a tuned amplifier having the frequency response of a single *LC* circuit

$$A = A_0 \frac{1}{1+Q^2(f/f_0 - f_0/f)^2}$$

and substitution into 6.16 yields

$$B_N = \frac{\pi f_0}{2Q} = \frac{\pi B}{2}$$ 6.17

where B is the bandwidth between the -3 dB points.

References

1. JOHNSON, J. B., *Phys. Rev.* **32**, 97 (1928)
2. NYQUIST, H., *Phys. Rev.* **32**, 110 (1928)
3. PIERCE, J. R., *Proc. Inst. Radio Engrs* **44**, 601 (1956)
4. JOLLY, W. P., *Low Noise Electronics* (E.U.P., London, 1967)
5. THOMPSON, B. J., NORTH, D. O., and HARRIS, W. A., *RCA Rev.* (1940–42).

Further Reading

ROBINSON, F. N. H., *Noise in Electrical Circuits* (O.U.P., London, 1962)
BENNETT, W. R., *Electrical Noise* (McGraw-Hill, New York 1960),
FREEMAN, J. J., *Principles of Noise* (Wiley, New York, 1958)
KING, R., *Electrical Noise* (Chapman and Hall, London, 1966)

PROBLEMS

1. Show that the noise bandwidth of the circuit in the diagram is $1/4CR$ Hz.

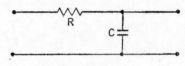

Fig. 6.11.

2. By considering the thermal noise e.m.f.'s in resistors R_1 and R_2 show that the r.m.s. noise voltage when these are connected in parallel is the same as that of the equivalent parallel resistor $R_1 R_2/(R_1 + R_2)$.

3. Find the r.m.s. noise across the 1 kΩ resistor in the diagram, assuming the diode to be saturated, and the observation bandwidth to be 10 kHz.

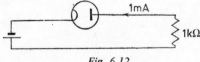

Fig. 6.12.

4. Repeat Problem 3, assuming the diode to be space-charge limited, the dynamic resistance being 5 kΩ and the cathode temperature 1 000°K.

5. Show the r.m.s. noise voltage across C is $\sqrt{(kT/C)}$.

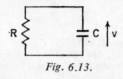

Fig. 6.13.

Chapter 7

Signals with Noise

7.1 INTRODUCTION

In Chapters 3 and 4 a linear relationship between bandwidth and signalling speed was established. The wider the bandwidth of a communication channel the larger the number of pulses which can be transmitted per second or the larger the number of speech or television channels which can be accomodated. But precision is just as important as speed. Unavoidably, noise is present on the channel, precision is lost, and there is an element of uncertainty about the received message. Hence the rate at which information can be sent depends both on the bandwidth and on the extent to which the signal power overrides the noise. An account of the numerical relationship among these three quantities must be delayed until we have a quantitative measure for information and this will be the concern of a later chapter. In this chapter we will be concerned with a re-examination of the modulation methods discussed in Chapter 4 from the point of view of signal-to-noise ratio and also with a closer examination of binary signalling systems in which one of two known signals is presented against a background of noise.

The scheme of the chapter is as follows. In Part A a mathematical model for noise waveforms is developed. Part B deals with the process of demodulation in the presence of noise. In Part C the extraction of signals of known waveform is discussed, together with the problem of deciding whether 'mark' or 'space' was transmitted when binary signalling is undertaken in the presence of noise.

205

A. A Mathematical Model for Noise Waveforms

7.2 NOISE STATISTICS

The first characteristic feature of most noise signals is that they are Gaussian. Electrical noise fluctuations are always the result of the superposition of a large number of tiny independent disturbances. It may therefore be argued, from the central limit theorem, that the value of the fluctuation observed at a given moment will be the value of a random variable having a Gaussian probability function. This means that if we observe the noise voltage, for example, across a resistor, at some moment, t_1, we may assume that the number obtained is a value of a Gaussian random variable. Indeed, there is good experimental evidence that, for thermal and other noise sources, this is the case.[1, 2]

It is tempting to go one step further and say that if a noise voltage is observed at times t_1, t_2, t_3, ... then a series of values of independent Gaussian variables is obtained. A glance at Fig. 6.1 however, shows that if $t_2 - t_1$, $t_3 - t_2$, etc., are sufficiently small then the result is a series of almost identical numbers. We suspect that successive measurements become independent if the time intervals are sufficiently long, but how long is 'sufficiently long'? We shall return to this question in the next section; for the moment we may avoid it by imagining a large number of observers, each equipped with a resistor of the same value and identical measuring equipment (or identically constructed noise diodes or noisy amplifiers). At a given moment each observer records the value of his noise signal. These measurements are surely independent and the fraction of measurements lying between v and $v + dv$ will be

$$p(v)\, dv = \frac{1}{\sqrt{(2\pi\sigma^2)}}\, e^{-v^2/2\sigma^2}\, dv \qquad\qquad 7.1$$

If each observer were to plot a graph of his noise waveform as a function of time then a collection of different noise waveforms of the type shown in Fig. 7.1 would result. Imaginary collections like this are very useful in statistics and such a collection is given the name *ensemble*. A single member of the ensemble is called a *sample function*. If we work across the ensemble at some given time, t^1, then values of a Gaussian random variable are obtained having variance σ^2 and zero mean (assuming that any d.c. component that

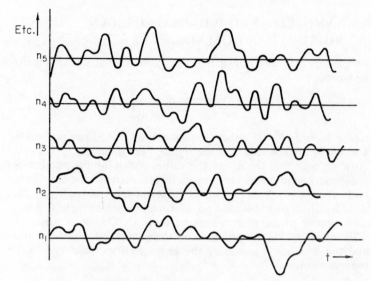

Etc.

n_5

n_4

n_3

n_2

n_1

$t \longrightarrow$

Fig. 7.1. Ensemble of random waveforms

might appear, as for example in observing shot noise, has been subtracted).

A second characteristic feature of noise fluctuations was noted in the previous chapter, namely that over a fairly wide frequency range, most noise sources are 'white', i.e. they have a uniform power spectrum. Usually it may be assumed that, within the relatively restricted bandwidth of a communication channel, the noise power spectrum is sensibly uniform and no harm comes from assuming that the power density is the same at frequencies outside the bandwidth, so long as the system is linear. That is to say, we assume that the noise power received per unit bandwidth is a constant, at least up to some upper frequency limit, F, which lies above the highest frequency to which the system will respond.

It should be stressed that Gaussian noise need not be white, neither need white noise be Gaussian. For our purposes, however, it will be sufficient to confine the discussion to a brief examination of interference by white Gaussian noise, band-limited to some upper angular frequency $\omega = 2\pi F$. In the next two sections a mathematical model for such noise signals is developed with a description first in the time domain, then in the frequency domain.

7.3 BAND-LIMITED WHITE GAUSSIAN NOISE—TIME DOMAIN

In Chapter 2 it was shown that any band-limited signal $n(t)$ may be written

$$n(t) = \sum_k n_k \text{ sinc } W\left(t - \frac{k\pi}{W}\right) \qquad 7.2$$

where $W = 2\pi F$, the upper frequency limit in radians per second and the n_k are the values of the signal at the moments $t_k = k\pi/W$. Now suppose that the n_k are the values of (an infinite number of) independent Gaussian random variables, each having zero mean and variance σ^2. A second set of values of the Gaussian variables will furnish us with a second waveform and, continuing in this way, an ensemble of waveforms may be built up. One such waveform is shown in Fig. 7.2. We proceed now to examine this model ensemble and see whether it possesses the characteristics described in the previous section.

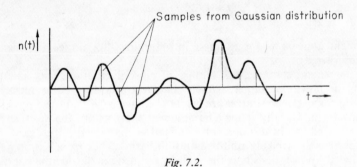

Fig. 7.2.

At any moment t', equation 7.2 shows that $n(t^1)$ may be written in the form

$$n(t') = \sum_k a_k n_k$$

where

$$a_k = \text{sinc } (Wt' - k\pi)$$

It follows that $n(t')$ is itself a random variable and, being the sum of Gaussian random variables, is itself Gaussian (see Section 5.12).

The mean and variance may be found from equations 5.23 and 5.24, thus

$$\overline{v(t')} = 0$$

$$\text{Var}\,[v(t')] = \sigma^2 \sum_k a_k^2$$

$$= \sigma^2 \sum_k \text{sinc}^2\,(Wt' - k\pi)$$

$$= \sigma^2$$

A method for evaluating the sum is indicated in Problem 11, Chapter 2.

We have now shown that the statistics running across this ensemble are Gaussian at any moment t' and not just at the sample moments t_k. The noise model therefore displays the first characteristic noted in the previous section. In order to show that our model also provides noise signals having a uniform power spectrum we first write down the autocorrelation function $R(s)$ for one of the sample functions of equation 7.2

$$R(s) = \lim_{T \to \infty} \frac{1}{T} \int_{-T/2}^{T/2} n(t)\,n(t+s)\,dt$$

$$= \lim_{T \to \infty} \left\{ \frac{1}{T} \int_{-T/2}^{T/2} \left[\sum_k n_k \,\text{sinc}\, W\left(t - \frac{k\pi}{W}\right) \right] \right. \qquad 7.3$$

$$\left. \times \left[\sum_l n_l \,\text{sinc}\, W\left(t + s - \frac{l\pi}{W}\right) \right] dt \right\}$$

when this is evaluated and averaged over the ensemble (see Appendix 4) the result is

$$R(s) = \sigma^2 \,\text{sinc}\,(Ws) \qquad 7.4$$

This is sketched in Fig. 7.3.

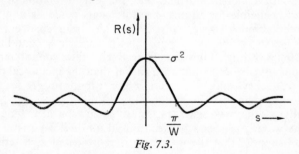

Fig. 7.3.

209

Four important results may be deduced from equation 7.4. First we note that $R(0) = \sigma^2$. Now $R(0)$, as we saw in Chapter 2, represents the mean power in the signal. It follows that the mean noise power is equal to the variance, hence

$$\overline{n^2} = \lim_{T \to \infty} \frac{1}{T} \int_{-T/2}^{T/2} n^2(t)\, \mathrm{d}t = \sigma^2 \qquad 7.5$$

The second result relates to the power spectrum, which is given by the Fourier transform of $R(s)$, hence

$$\begin{aligned} p(\omega) &= \sigma^2 \pi / W = \sigma^2 / 2F \quad &-W < \omega < W \\ &= 0 &|\omega| > W \end{aligned} \qquad 7.6$$

Thus the power spectrum is white whithin the band which is the second characteristic required. The units in equation 7.6 are such that the one-sided power density (see equation 2.60) is $p(\omega)/\pi V^2$ rad^{-1} s. For the remainder of this chapter we shall use the notation p_n to represent this quantity. From equation 7.6

$$p_n = \frac{\sigma^2}{W} = \frac{\overline{n^2}}{W} \qquad 7.7$$

The third result is the answer to our earlier question 'how long must we wait after measuring a noise signal before a second measurement gives a value independent of the first?' Fig. 7.3 shows that measurements made at intervals $1/2F$, i.e. the Nyquist sampling interval, are uncorrelated. The Gaussian property implies that they are independent. Again this applies to measurements made between the initially chosen sampling instants t_k (we arranged for the values at the moments t_k to be independent in the initial assumptions about the model). Further, since the autocorrelation function becomes negligible for large s, measurements made at any arbitrary interval much larger than $1/2F$ will also be independent. This result is not surprising if we suppose that the noise was band-limited by passing it through a low-pass filter. Such a filter cannot 'remember' input signals occurring much earlier than the reciprocal of the cut-off frequency, since its impulse response tends to zero for times much longer than this.

Finally, we may ask 'what fraction of some arbitrarily long time does the waveform spend between n and $n+\mathrm{d}n$?' To answer this question suppose that the waveform is represented as a series of adjacent, indefinitely short, rectangular impulses. Any set of impulses taken $1/2F$ apart have amplitudes which are statistically independent and hence the fraction of them which lie between n and $n+\mathrm{d}n$ is given by equation 7.1. But the same is true of the next adjacent set of impulses and so on until we have dealt with the whole waveform. It follows that the *fraction of the time* which the signal spends between n and $n+\mathrm{d}n$ is

$$f(n)\,\mathrm{d}n = \frac{1}{\sqrt{(2\pi\sigma^2)}}\,\mathrm{e}^{-n^2/2\sigma^2}\,\mathrm{d}n \qquad\qquad 7.8$$

More generally it follows that the average of any function of n will be the same whether evaluated in time or across the ensemble. Ensembles having this property are said to be *ergodic*. An example was met in equation 7.5 where the mean of n^2 in time (the mean power delivered to a 1 Ω resistor) turns out to be the same as the ensemble mean square, σ^2.

7.4 BAND-LIMITED WHITE GAUSSIAN NOISE—FREQUENCY DOMAIN

For many problems it is more convenient to describe a noise signal in the frequency domain. We may attempt to do this as follows, starting from equation 7.2, and cheating a little. Consider the central batch of samples n_k for which $-(WT/2\pi) < k < (WT/2\pi)$ i.e., those in the range $-T/2 < t_k < T/2$. Suppose we remove all the samples outside this range and replace them with batches identical to the central one, so that samples at intervals T are the same. It is clear that, with these sample values, equation 7.2 no longer describes a truly random waveform but one which repeats with period T (Fig. 7.4). If T is made longer than any observation time envisaged, then the repetition need be of no physical consequence.

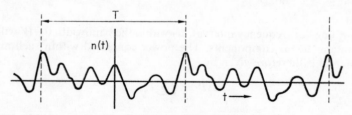

Fig. 7.4. 'Noise' waveform of equation 7.9

211

It means, however, that we may Fourier-analyse the waveform and obtain sine and cosine coefficients at the discrete angular frequencies which are multiples of $2\pi/T$.

The frequency analysis of this repetitive version of the signal of equation 7.2 is given in Appendix 5. The details are of no special interest but the results are useful and may be stated as follows. The noise waveform $n(t)$ which has been forced to repeat with period T may be expanded in the form

$$n(t) = a_0 + \sum_{m=1}^{WT/2\pi} a_m \cos m\omega_1 t - \sum_{m=1}^{WT/2\pi} b_m \sin m\omega_1 t \qquad 7.9$$

where a_0, a_m, and b_m are values of independent Gaussian random variables of zero mean. The mean square value of the a_m and b_m coefficients is

$$\overline{a_m^2} = \overline{b_m^2} = p_n \omega_1 \qquad 7.10$$

where p_n is given by equation 7.7. The d.c. term has mean square

$$\overline{a_0^2} = \frac{p_n \omega_1}{2} = \frac{\overline{n^2}}{N} \qquad 7.11$$

where N is the total number of independent samples in the interval T. Since a_0 is the mean value of the waveform, and therefore the mean value of the samples (Section 2.16), this last result might have been anticipated from equation 5.26.

Equation 7.9 may also be expressed in the form

$$n(t) = a_0 + \sum_{m=1}^{WT/2\pi} c_m \cos (m\omega_1 t - \phi_m) \qquad 7.12$$

where $$c_m^2 = a_m^2 + b_m^2$$

and $$\tan \phi_m = \frac{b_m}{a_m}$$

Any angular frequency interval $\Delta\omega$ within the bandwidth $0 - W$ will contain $\Delta\omega/\omega_1$ components. The power contained within such an interval is therefore given by

$$\Delta P = \tfrac{1}{2} \sum_m c_m^2$$

where the summation includes only those components within $\Delta\omega$.

If the number of components is large then, by the law of large numbers, ΔP is certain to lie very close to

$$\Delta P = \frac{1}{2} \frac{\Delta \omega}{\omega_1} \overline{c_m^2} = \frac{1}{2} \frac{\Delta \omega}{\omega_1} (\overline{a_m^2 + b_m^2})$$

$$= p_n \Delta \omega$$

Thus, provided T is sufficiently large (i.e. $2\pi/T = \omega_1 \ll \Delta \omega$) the power density is constant, i.e. $(\Delta P/\Delta \omega) = p_n$ in agreement with the analysis of the previous section.

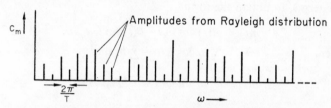

Fig. 7.5. *Amplitude spectrum of waveform of Fig. 7.4*

The spectrum of this noise signal has the form sketched in Fig. 7.5. The reader may have already observed that the random numbers c_m are not Gaussian but have a Rayleigh distribution (Section 5.12). The phase angles ϕ_m are quite randomly distributed around 2π.

7.5 NARROWBAND GAUSSIAN NOISE

This type of noise results when white Gaussian noise is passed through a bandpass filter of relatively narrow bandwidth. It will accompany a carrier which has been filtered and amplified by the i.f. stages of a receiver. The results of this section will therefore be especially valuable when the demodulation of such signals is considered in Part B of this chapter. A useful description starts from equation 7.9,

$$n(t) = \sum_m a_m \cos m\omega_1 t - \sum_m b \sin m\omega_1 t$$

in which the range of the integer m is sufficient to include only those frequencies $m\omega_1$, which lie within the bandwidth β of the filter whose centre frequency is ω_c (assumed to be an integer multiple of ω_1). If we define an integer l as

$$l = m - \omega_c/\omega_1$$

then
$$-\beta/2\omega_1 < l < \beta/2\omega_1.$$

We may therefore write

$$n(t) = \sum_l a_l \cos(l\omega_1 + \omega_c)t - \sum_l b_l \sin(l\omega_1 + \omega_c)t \qquad 7.13$$

It is helpful to rewrite this expression, expanding the cosine and sine terms, to obtain the form

$$n(t) = A(t) \cos \omega_c t - B(t) \sin \omega_c t \qquad 7.14$$

where
$$A(t) = \sum_l a_l \cos l\omega_1 t - \sum_l b_l \sin l\omega_1 t$$

and
$$B(t) = \sum_l a_l \sin l\omega_1 t + \sum_l b_l \cos l\omega_1 t$$

We pause at this point to note certain properties of $A(t)$ and $B(t)$. First, because of the assumption of narrow bandwidth, $l\omega_1 \ll \omega_c$ for all l so that the functions $A(t)$ and $B(t)$ vary only slowly compared with $\cos \omega_c t$ and $\sin \omega_c t$. Further, they are linear combinations of the Gaussian random variables a_l and b_l and therefore are themselves Gaussian random variables. Although knowledge of the a_l and b_l completely determines both $A(t)$ and $B(t)$, a knowledge of $A(t^1)$ at a given moment, t^1, does not give any knowledge of $B(t^1)$. It is not difficult to show that, averaging across the ensemble, the expectation of the product $A(t^1) \times B(t^1)$ is zero, thus these random Gaussian variables are uncorrelated and independent. Reference to equation 7.10 shows that $A(t)$ and $B(t)$ may each be regarded as a white noise signal with mean square

$$\sigma_N^2 = \overline{A(t)^2} = \overline{B(t)^2} = \beta p_n \qquad 7.15$$

having uniform power density, $2p_n$, from d.c. to $\beta/2$. The factor 2 arises because the sine and cosine coefficients for $A(t)$ and $B(t)$ contain a_l and a_{-l}, b_l and b_{-l}, respectively. If the functions A and B were constants, the mean power of the waveform of equation 7.14 would be $(A^2 + B^2)/2$. The values for $\overline{A(t)^2}$ and $\overline{B(t)^2}$ noted above imply a mean power βp_n which is exactly the power transmitted by the filter of bandwidth β.

214

Finally, equation 7.14 is rewritten in the form

$$n(t) = E(t) \cos [\omega_c t - \phi(t)] \qquad 7.16$$

where $$E(t) = \sqrt{[A(t)^2 + B(t)^2]}$$

and $$\tan \phi(t) = \frac{B(t)}{A(t)}$$

The statistics of $E(t)$ and $\phi(t)$, looking across the ensemble, are such that $\phi(t)$ is random in the range 0 to 2π while $E(t)$ is recognised as having a Rayleigh distribution of the form

$$p(E) = \frac{E}{\sigma_N^2} e^{-E^2/2\sigma_N^2} \qquad 7.17$$

This function is shown in Fig. 7.6. The mean and variance of this distribution were evaluated in Chapter 5; the same result would be obtained by averaging in the time domain so we have

$$\overline{E(t)} = \sqrt{\left(\frac{\pi \sigma_N^2}{2}\right)}$$

$$\overline{E(t)^2} = 2\sigma_N^2 \qquad 7.18$$

$$\sigma_E^2 = \overline{E(t)^2} - \overline{E(t)}^2 = (2 - \pi/2)\sigma_N^2$$

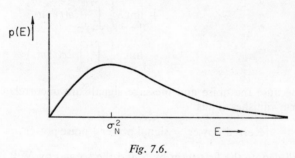

Fig. 7.6.

where $\sigma_N^2 = \beta p_n$ is the mean power of the narrowband noise signal, filtered from a white noise signal of power density p_n.

To recapitulate, equation 7.16 shows that a narrowband noise signal behaves like a carrier $\cos \omega_c t$ which suffers amplitude and phase modulation in a random fashion. Fig. 6.1(d) illustrates such

Signals and Information

a signal, although the bandwidth in that example is not particularly narrow. A radio receiver tuned to ω_c in the absence of a carrier will present a signal having the form of equation 7.16 to the demodulator. If an envelope detector is employed, $E(t)$ is the signal passed on to the audio amplifier stages. The mean power is given by equation 7.18. Our rather special use of power, as mean square voltage or current, should be borne in mind; no violation of energy conservation is implied in equation 7.18.

B. Signal-to-Noise Ratios Following Demodulation

7.6 SIGNAL-TO-NOISE RATIO

At the receiving end of a communication channel, the observed waveform $f(t)$ will consist of the message signal $v(t)$, which may be either a direct analogue signal or a carrier modulated by one of the methods described in Chapter 4, together with additive noise $n(t)$, thus

$$f(t) = v(t) + n(t) \qquad 7.19$$

The mean power in the received signal is given by

$$\lim_{T \to \infty} \frac{1}{T} \int_{-T/2}^{T/2} f(t)^2 \, dt = \lim_{T \to \infty} \frac{1}{T} \int_{-T/2}^{T/2} v(t)^2 \, dt$$
$$+ \lim_{T \to \infty} \frac{1}{T} \int_{-T/2}^{T/2} n(t)^2 \, dt$$
$$+ \lim_{T \to \infty} \frac{1}{T} \int_{-T/2}^{T/2} 2v(t)\, n(t) \, dt$$

and, because the noise and message signals are uncorrelated, the last term vanishes. Thus

received power = signal power + noise power

We will denote the first term by S and the second by N. It is clear that here we may regard the signal-to-noise ratio S/N as *either* 'the received signal power in the absence of noise divided by the received noise power in the absence of a signal' *or* 'the ratio of the contributions to the total power by signal and noise, respectively'.

It must be emphasised, however, that these definitions are only identical because of the linear superposition expressed by equation

216

7.19. Received signals are always subject to non-linear processing during frequency-changing and demodulation, so that special care is needed in interpreting the term signal-to-noise ratio in these circumstances. In the process of frequency changing, the waveform of equation 7.19 may be added to a locally supplied sinusoid and presented to a non-linear device. The output will then contain terms due to the local oscillator beating with the signal and beating with the noise as well as terms due to signal beating with noise, signal with signal and noise with noise (both considered broken down into sinewave components). Provided the local oscillator signal is sufficiently strong—and as explained in Chapter 3 this is always arranged to be the case—the latter group of terms become second-order and may be neglected. The process is then essentially a linear one, transposing the whole signal up or down the frequency axis. This leaves signal-to-noise ratios unaffected.

When an a.m. signal whose carrier power sufficiently exceeds that of the noise is demodulated by a non-linear device, similar considerations apply. In this case it is more convenient to use the first definition given above but to interpret 'signal' to denote the presence of sidebands and to assume that the carrier is always present. The 'signal power' measured after demodulation then ignores any d.c. component due to the carrier. Similar considerations apply to the demodulation of f.m. signals. In Sections 7.8 and 7.9 noise and signal will be treated separately, *but assuming the presence of a strong carrier*. This is always appropriate for broadcast speech or music signals in which the amount of noise present during quiet passages or pauses is of most significance, since it is then most irritating. The more general analysis of signal-to-noise ratios when the carrier is weak is much more involved and will not be treated here.

7.7 SYNCHRONOUS DEMODULATION OF A. M. SIGNALS IN NOISE

In Fig. 7.7 the group of solid lines represents the signal+noise which would be presented to the demodulator of an s.s.b. receiver tuned to a carrier at ω_c and with a pass band extending from ω_c to $\omega_c+\omega_b$. The modulating signal is a sinewave of frequency ω_m. It is assumed that the upper sideband is transmitted, thus

$$f(t) = a \cos(\omega_c+\omega_m)t+n(t) \qquad 7.20$$

where $n(t)$ is the repetitive noise signal discussed in Sections 7.4

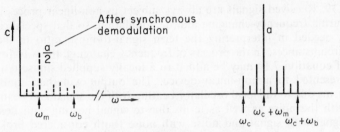

Fig. 7.7. S.S.B. signal in noise

and 7.5. With white noise of spectral density p_n (equation 7.7), the noise component $n(t)$ has a power $p_n\omega_b$ while the signal power is $a^2/2$.

A synchronous demodulator processes $f(t)$ by multiplying it with a locally supplied signal $\cos \omega_c t$, synchronous with the carrier, and passing on the difference terms. The demodulated output due to the signal sideband is thus $(a/2) \cos \omega_m t$ and is shown in Fig. 7.7 as a dotted line at ω_m. The noise 'sidebands' will also be shifted down the frequency scale, to lie between d.c. and ω_b, and similarly reduced in amplitude by a factor of two. As both signal and noise power have been reduced in power by a factor of four, *the signal-to-noise ratio is unchanged*. We may write

$$\left(\frac{S}{N}\right)_{\text{s.s.b.}} = \frac{a^2}{2p_n\omega_b} = \left(\frac{S}{N}\right)_D \qquad 7.21$$

where $(S/N)_D$ is the signal-to-noise ratio which would be achieved if the same signal power were used for direct analogue transmission in the baseband, d.c. to ω_b, in the presence of noise having the same spectral density as that accompanying the carrier. This establishes a useful reference (S/N) level for comparing different modulation systems.

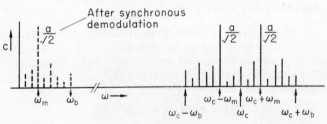

Fig. 7.8. D.S.B. signal in noise

Signals and Noise

A d.s.b. signal is illustrated in Fig. 7.8. This may be written

$$f(t) = \frac{a}{\sqrt{2}} \cos (\omega_c + \omega_m)t + \frac{a}{\sqrt{2}} \cos (\omega_c - \omega_m)t + n^1(t) \qquad 7.22$$

in which we have arranged to maintain the total transmitted power at $a^2/2$ to enable a direct comparison with s.s.b. to be made. The noise power in $n^1(t)$ is clearly twice that in 7.19 because the bandwidth is now $\beta = 2\omega_b$. Multiplying equation 7.22 by $\cos \omega_c t$ and picking out the difference terms, we find two signal terms which add coherently to give a demodulated signal $(a/\sqrt{2}) \cos \omega_m t$. The noise in the baseband, d.c. to ω_b, now consists of two superimposed batches of noise 'sidebands', one which ran from ω_c to $\omega_c + \omega_b$ and the other which ran from ω_c to $\omega_c - \omega_b$. As in the s.s.b. case, each batch is reduced in power by a factor of four and hence (recalling that noise powers add) the noise power after demodulation is twice that of the s.s.b. system. The signal power after demodulation is also twice that of the s.s.b. system. Hence

$$\left(\frac{S}{N}\right)_{\text{d.s.b.}} = \left(\frac{S}{N}\right)_{\text{s.s.b.}} = \left(\frac{S}{N}\right)_D \qquad 7.23$$

when the same power is transmitted in the sidebands and the appropriate receiver bandwidths are used.

We note, in passing, that the result obtained here, namely that s.s.b. or d.s.b. offers no advantage over direct transmission in the baseband, assumes that the added noise is white. If the noise is not white then modulation may be used to translate the information into a less noisy part of the spectrum before transmission. In the previous chapter we noted that amplifiers tend to be particularly noisy at low frequencies due to $1/f$ or 'flicker' noise. When d.c. or very-low-frequency signals are to be measured, a distinct advantage can often be achieved if the signal can be d.s.b. modulated (simply switched on and off) at a frequency where low noise amplification can be obtained. The amplified signal is then synchronously demodulated. This is the 'phase-sensitive detector' method, a technique commonly used for the measurement of low-frequency or d.c. signals. We shall return to this in Section 7.16.

The d.s.b. signal just considered may be changed to a normal a.m. signal by adding a carrier term $\cos \omega_c t$ of sufficient amplitude. The demodulated output will be unchanged apart from a d.c. term which carries no information. Thus, provided we maintain

219

the same power in the sidebands, there is no difference in the signal-to-noise ratios after a synchronous demodulator whether s.s.b., d.s.b., or conventional a.m. modulation is employed.

These remarks underline the basic similarity between the a.m. systems but, while the results are of academic interest, *they do not relate to the best possible signal-to-noise performance in a practical*

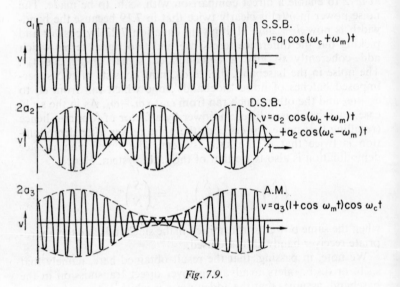

Fig. 7.9.

situation. For in practice it is the peak signal amplitude which can be transmitted, before overload or non-linearity sets in, which determines the maximum signal power and thus the best signal-to-noise ratio which can be achieved. It is here that the difference between the modulation systems shows up. Fig. 7.9 shows the waveforms for s.s.b., d.s.b., and a fully modulated a.m. signal. With the notation of that figure, it is easily shown that the peak amplitudes will be the same if

$$a_1 = 2a_2 = 2a_3 \qquad\qquad 7.24$$

The sideband powers are in the ratio

$$\frac{a_1^2}{2} : a_2^2 : \frac{a_3^2}{4}$$

and expressing each of these in terms of a_1 using equation 7.24 the sideband powers become

$$\frac{a_1^2}{2} : \frac{a_1^2}{4} : \frac{a_1^2}{16}$$

or

$$8 : 4 : 1$$

The relative sideband powers under peak power limitation are thus related as 9 dB, 6 dB and 0 dB for the s.s.b., d.s.b., and a.m. signals, respectively. It follows that if each of these signals were received in the presence of the same white noise background by receivers of appropriate bandwidth and employing synchronous demodulators, the output signal-to-noise ratios would be related in the same way. This is the famous result, that s.s.b. offers a 9 dB superiority over conventional a.m., and 3 dB over d.s.b. We shall show in the next section that envelope demodulation of a.m. offers the same signal-to-noise performance as synchronous demodulation, provided that the carrier is of sufficient amplitude. It must be noted, however, that the 9 dB superiority of s.s.b. was calculated for sinusoidal test tone modulation and the same figure need not apply with other modulating signals.

7.8 ENVELOPE DEMODULATION OF A STRONG CARRIER IN NOISE

A general analysis of the envelope detection of a noisy a.m. signal is difficult and will not be attempted here. The discussion is restricted to the case of large signal-to-noise ratios with an analysis following closely that given in Chapter 4 for interference by a neighbouring carrier.

It was shown in Chapter 4 that amplitude and angle modulation of a sinusoidal carrier are conveniently described by means of a phasor diagram, viewed from a reference frame rotating with the carrier angular frequency ω_c. Fig. 7.10 shows the phasor diagram for an unmodulated carrier in the presence of a noise signal which has been filtered by the preceding i.f. stages of bandwidth β. The carrier, $c \cos \omega_c t$, is represented by a horizontal phasor of length c and to this are added the phasors representing the narrowband noise signal in the form described by equation 7.14. The resultant phasor R has magnitude

$$R(t) = \sqrt{\{[c + A(t)]^2 + B(t)^2\}}$$

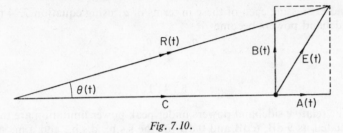

Fig. 7.10.

and angle
$$\theta(t) = \tan^{-1} \frac{B(t)}{c + A(t)}$$

The end point of R wanders randomly about the undisturbed position. Assuming a large signal-to-noise ratio, or more precisely if c exceeds the r.m.s. value of $A(t)$ or $B(t)$ by at least an order of magnitude (corresponding to a carrier-power-to-noise-power ratio exceeding about 17 dB), the above expressions may be approximated as

$$R(t) \approx c + A(t) \qquad\qquad 7.25$$

$$\theta(t) \approx \frac{B(t)}{c} \qquad\qquad 7.26$$

An envelope detector will ignore the phase variation and give as its output the signal described by equation 7.25. It contains a d.c. component (which we ignore) together with $A(t)$ which as we have seen in Section 7.5 is a Gaussian noise signal whose spectrum is white with spectral density $2p_n$ up to a band limit $\beta/2$. The output noise power in the absence of modulation is thus precisely the noise power βp_n received in the i.f. bandwidth (multiplied, of course, by the gain of the i.f. amplifiers).

To make a direct comparison with the synchronous demodulation process, let us for convenience suppose the noise to be absent and add—as in the previous section—a pair of sidebands, each of amplitude $a/\sqrt{2}$. If $c > a/\sqrt{2}$ we now have an a.m. signal of modulation depth $a\sqrt{2}/c$. The envelope detector produces a sine wave of amplitude $a\sqrt{2}$ and hence the output signal power $S = a^2$. The output signal-to-noise power ratio is thus

$$\frac{S}{N} = \frac{a^2}{\beta p_n} = \frac{a^2}{2\omega_b p_n}$$

which is identical to that obtained using synchronous detection (equations 7.21 and 7.23) with the same input signal. For reference we write the signal-to-noise power obtained with either variety of demodulator, when presented with a fully modulated strong carrier, in terms of the carrier amplitude c

$$\left(\frac{S}{N}\right)_{a.m.} = \frac{c^2}{4\omega_b p_n} \qquad 7.27$$

It is natural to enquire how it comes about that the envelope detector performs like a synchronous detector at large signal-to-noise ratios. The operation of a diode detector, however, may be regarded as synchronous under these circumstances, for a strong carrier dominating the signal will switch the diode in much the same way as happens in the process of synchronous detection.

At low carrier-to-noise ratios an envelope detector offers a poorer performance than a synchronous detector. As we saw in Section 7.5, when a carrier is totally absent, the output signal will be $E(t)$. This fluctuates about a mean value which is not zero and we equate, as usual, noise power to the variance σ_E^2

$$\sigma_E^2 = (2-\pi/2)\sigma_N^2 = (2-\pi/2)\beta p_n$$
$$= 0.429\beta p_n$$

This is less than the output noise in the presence of a strong carrier. The appearance of a carrier thus increases the output noise and this results in a poorer signal-to-noise performance than for synchronous detection at low carrier-to-noise levels.

7.9 FREQUENCY DEMODULATION OF A STRONG CARRIER IN NOISE

Now let the resultant phasor of the previous section be applied to a discriminator which gives a voltage output

$$v = D\frac{d\theta}{dt}$$

where $d\theta/dt$ is the instantaneous frequency deviation and D is a constant of the discriminator. From equation 7.26 we may write

$$v(t) = \frac{D}{c}\frac{dB(t)}{dt}$$

223

The function $B(t)$ is also a white Gaussian noise signal whose sine and cosine Fourier coefficients have mean square values $2p_n\omega_1$ (equation 7.15 and Section 7.5). Now the process of differentiation, above, introduces an additional factor ω into the amplitude of each Fourier component of $v(t)$ at frequency ω. Hence, at frequency $\omega = l\omega_1$ the sine and cosine coefficients of this signal each have mean square values

$$\frac{2D^2 p_n l^2 \omega_1^3}{c^2}$$

so that the output noise power density is

$$p_o(\omega) = \frac{2D^2 p_n \omega^2}{c^2} \qquad 7.28$$

This is illustrated in Fig. 7.11.

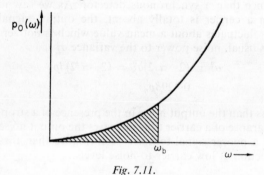

Fig. 7.11.

Only that part of the noise spectrum from d.c. to ω_b will be passed by the subsequent amplifiers, hence the effective *total* output noise power is

$$N = \frac{2D^2 p_n}{c^2} \int_0^{\omega_b} \omega^2 \, d\omega$$

$$= \frac{2D^2 \omega_b^3 p_n}{3c^2} \qquad 7.29$$

The above expression gives the output noise power in the presence of an unmodulated carrier. As in the previous section (see the discussion in Section 7.6) we now compare this with the output signal power when the carrier is fully modulated using a sinusoidal test tone which introduces a peak frequency deviation $\Delta\omega$ which is the maximum permitted by the design of the system. The discriminator now gives a sine wave output signal of amplitude $D\Delta\omega$ and hence the output signal power is

$$S = \frac{D^2(\Delta\omega)^2}{2}$$

Hence

$$\left(\frac{S}{N}\right)_{\text{f.m.}} = \frac{3c^2(\Delta\omega)^2}{4\omega_b^3 p_n}$$

On substituting $\Delta\omega/\omega_b = \beta_b$ (the modulation index for the highest modulating signal frequency permitted) this becomes

$$\left(\frac{S}{N}\right)_{\text{f.m.}} = \frac{3c^2\beta_b^2}{4\omega_b p_n} = 3\beta_b^2\left(\frac{S}{N}\right)_{\text{a.m.}} \qquad 7.30$$

where equation 7.27 has been used.

With $\beta_b = 5$, which is typical for a commercial wideband f.m. system, this represents a very substantial improvement of about 19 dB over an a.m. system. A further improvement may be achieved with the additional sophistication of pre-emphasis and de-emphasis techniques, but the reader is referred elsewhere for a treatment of these (see, for example, Schwartz[3]).

This improvement over a.m. (and direct baseband transmission) has not been won without cost. The price paid is the wider bandwidth occupied by the f.m. signal. The larger β_b the wider the bandwidth but the better the signal-to-noise ratio—or, for a specified signal-to-noise ratio, the lower the transmitter power needed. Frequency modulation is an example of a modulation system which enables a kind of exchange of signal-to-noise ratio for bandwidth to be made. It furnishes a flexibility in design which is not possible with a.m. If the channel is noisy, the noise may be overcome by increasing the bandwidth of the signal. This means, of course, that fewer channels can be accomodated in a given part of the spectrum. Noise always reduces the rate at which information can be transmitted using a given transmitter power.

If the interfering noise power is increased sufficiently, the analysis given above ceases to be valid, for the value of $E(t)$ in Fig. 7.10 will no longer be small in comparison with the carrier amplitude. In a wideband f.m. system the modulating signal swings the carrier phasor through several revolutions, whereas an interfering signal which is smaller in amplitude than the carrier can never accomplish phase changes in excess of $\pi/2$. As soon as $E(t) > c$, however, the resultant $R(t)$ can suffer large phase displacements, determined by $E(t)$. The noise takes over. This gives rise to a *threshold effect* common, as we shall see, to other wideband improvement systems. Above a certain carrier level, f.m. enjoys the signal-to-noise superiority over synchronously demodulated a.m. expressed by equation 7.30, while below this threshold the performance rapidly deteriorates, as shown in Fig. 7.12, eventually becoming worse than for a.m.

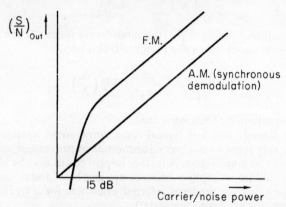

Fig. 7.12. Threshold effect in wideband f.m.

7.10 UNCODED PULSE TRAIN CARRIERS IN NOISE

In the previous three sections the influence of noise on modulated sinusoidal carriers has been studied and the various systems compared on the basis of the signal-to-noise ratio obtaining at the demodulator output. Space does not permit any extended study of pulse modulation systems and the present section is largely restricted to a discussion of pulse amplitude modulation.

Consider a multiplexed p.a.m. system transmitting M signals, each band-limited to ω_b. Pulses at a rate ω_b/π are required for each signal so that there will be a total of $M\omega_b/\pi$ pulses per second on the channel. We have seen in Chapter 4 that, to avoid intersymbol interference and to keep the channel bandwidth to a minimum, an ideal system would employ pulses of the form sinc $M\omega_b t$ and thus the minimum channel bandwidth is $\omega_c = M\omega_b$.

Suppose that one of the signals is a sinusoid so that its pulses have amplitudes varying as a sine wave of amplitude a. We first need to know the signal power P_s required to do this. (P_s is the signal power on the channel, and is not necessarily the same as the signal power after demodulation denoted by S.) The energy in a single pulse, v sinc $\omega_c t$, is $\pi v^2/\omega_c$ (Section 2.16). For our sine wave test signal $\overline{v^2} = \overline{a^2}/2$ and these pulses occur at ω_b/π per second hence

$$P_s = \frac{\pi a^2}{2\omega_c} \times \frac{\omega_b}{\pi} = \frac{a^2\omega_b}{2\omega_c} \qquad 7.31$$

This result might have been anticipated, for the fraction of the time being allotted to the sine wave is $1/M$ or ω_b/ω_c.

Demodulation proceeds by sampling the received waveform at a rate ω_b/π in synchronism with the desired signal samples. The mean square sample height will be expressed as the sum of two terms, one due to the signal and one due to the noise, these being uncorrelated. The variance resulting from the noise equals the total noise power $p_n\omega_c$. Hence the signal-to-noise ratio after demodulation is

$$\frac{S}{N} = \frac{a^2}{2p_n\omega_c}$$

On substituting for a^2 from equation 7.31 we obtain

$$\left(\frac{S}{N}\right)_{\text{a.m.}} = \frac{P_s}{p_n\omega_b} = \left(\frac{S}{N}\right)_D \qquad 7.32$$

It follows that the same signal-to-noise performance would be achieved by direct transmission of the analogue signal in the baseband (compare equation 7.21). The same result was found for s.s.b. transmissions. Time-division-multiplexed p.a.m. and frequency-division-multiplexed s.s.b. also require (ideally) the same channel bandwidth. A choice between them would therefore be governed by other factors such as cost, freedom from cross-talk, etc.

A quite different situation arises with other pulse modulation systems such as pulse position modulation, p.p.m. Fig. 7.13 illustrates the principle of the method. The signal crosses some voltage

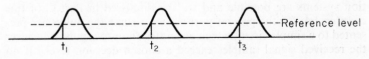

Fig. 7.13.

reference level at the moments t_1, t_2, t_3, etc., which are displaced from those for an unmodulated pulse train by an amount proportional to the sample values of the modulating signal. The influence of a *small* added noise signal may be thought of qualitatively as lifting each pulse up (or down) by a small random amount. Because of the finite rise time of the pulses this results in an error in time proportional to the vertical displacement. If the channel bandwidth is increased, the pulse rise times may be proportionately decreased whereas the r.m.s. noise level which determines the displacements increases only as the square root of the bandwidth. These two effects give an r.m.s. error inversely proportional to the square root of the bandwidth and hence an output noise component inversely proportional to the bandwidth. Finally, we recall that, as above, the power on the channel required for pulses of given amplitude varies as the reciprocal of the channel bandwidth. The net result is thus a signal-to-noise ratio which improves as the square of the bandwidth. This is similar to the result for f.m; indeed, if β_m is sufficiently large, equation 7.30 also shows an improvement varying as the square of the bandwidth. P.P.M. also exhibits a 'threshold effect' similar to that encountered for f.m.

7.11 NOISE IN PULSE CODE MODULATION. QUANTISATION NOISE

A discussion of noise in p.c.m. systems falls naturally into two parts. First, there is the quantisation noise introduced in the process of coding prior to transmission; this has been mentioned in Chapter 4. Second, there is the influence of Gaussian noise in the channel which will give rise to errors in the 'pulse' or 'no pulse' decisions made at the receiving end. It is more convenient to begin with the latter.

We have seen, in Chapter 4, that the bandwidth needed for the transmission of a binary coded signal having $L = 2^l$ quantisation levels is $l\omega_b$ where ω_b is the baseband-limiting angular frequency. If the transmission is received in the presence of white Gaussian noise having power spectral density p_n, the variance of a noise sample voltage will be $\sigma^2 = l\omega_b p_n$. Now, suppose the pulse height is h and the decision level is set at $h/2$. (More sophisticated pulse detection systems are possible and will be discussed in Part C of this chapter, but for the present purpose we assume the pulses are presented to a simple circuit which generates an output pulse whenever the received signal samples exceed a chosen decision level.) If no pulse is present at the sampling instant, then an error will be made should the noise voltage happen to exceed $h/2$ at that instant. If a pulse is present then an error will be made should the noise voltage happen to be less than $-h/2$. In either case the probability of error is given by

$$P(\varepsilon) = \int_{h/2}^{\infty} \frac{1}{\sqrt{(2\pi\sigma^2)}} \, e^{-x^2/2\sigma^2} \, dx \qquad 7.33$$

This may be related to the integrals encountered in the discussion of the Gaussian probability distribution in Section 5.12. From equation 5.35 we have

$$P(\varepsilon) = \frac{1}{2} \left[1 - \text{erf}\left(\frac{h}{2\sqrt{2\sigma}} \right) \right]$$

The signal has sample values which are either h or zero, with roughly equal probability. It is left as an exercise for the reader to show that the signal power on the channel, P_s, is $h^2/2$. Setting the noise power on the channel as $N_c = \sigma^2$, the last equation becomes

$$P(\varepsilon) = \frac{1}{2} \left[1 - \text{erf}\left(\frac{1}{2} \sqrt{\frac{P_s}{N_c}} \right) \right] \qquad 7.34$$

Some numerical values are given in Table 7.1, in which the last column shows the mean time interval between errors in a typical system which employs eight-binit words and a baseband extending to 4 kHz. There is clearly a signal power level, at around 20 dB above the noise, above which errors become negligible. The pulse train passed on by the decision circuits in the receiver (or repeater

TABLE 7.1

Pulse height: r. m. s. noise (dB)	P_s/N_c (dB)	$P(\varepsilon)$	τ
8	5	$0 \cdot 105$	150 μs
11	8	$3 \cdot 8 . 10^{-2}$	410 μs
14	11	$6 \cdot 1 . 10^{-3}$	2·5 ms
17	14	$2 \cdot 0 . 10^{-4}$	78 ms
20	17	$2 \cdot 8 . 10^{-7}$	56 s
23	20	$7 \cdot 7 . 10^{-13}$	240 d
26	23	$7 \cdot 6 . 10^{-24}$	$6 \cdot 7 . 10^{10}$ a

229

station) is then identical to that sent from the transmitter. The probability that this pulse train will differ significantly from that produced immediately after coding can be made as small as we please by using a signal power which exceeds the noise by a modest 20 dB or so. This will remain the case even after a sequence of repeaters mas distinct from analogue modulation where the noise received is also amplified and passed on. Furthermore, the repeaters are

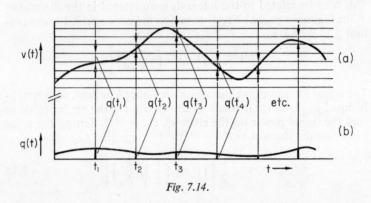

Fig. 7.14.

here pulse generators and no stringent linearity specifications need be met. The only remaining noise on the demodulated signal is the quantisation noise and this is under the control of the designer, as we shall now see. This is the great attraction of p.c.m.

Fig. 7.14(a) illustrates the process of quantisation already described in Chapter 4. The errors introduced in this process are clearly identical to those which would be obtained if the signal of Fig. 7.14(b) were to be added to the signal prior to coding. This error signal $q(t)$ has been constructed using the usual sinc $\omega_b t$ interpolation procedure and taking as samples the error voltages q_k, as indicated. The argument used in Section 7.3 may be used to show that the power spectrum of $q(t)$ is uniform up to the band limit ω_b. It is also plausible to assume that the probability density function across an ensemble for $a(t)$ will be the rectangular one given in Problem 3 of Chapter 5. The variance of this distribution is easily shown to be $Q^2/12$, where Q is the quantisation interval. The mean quantisation noise power is

$$N_Q = \sigma_Q^2 = Q^2/12$$

If the signal were a sine wave test tone, an amplitude $QL/2$ would

230

fully modulate the carrier, hence the signal power may be written

$$S = Q^2L^2/8$$

The mean square signal-to-noise power may therefore be written

$$\frac{S}{N_Q} = 10 \log_{10} \frac{3L^2}{2} = 10 \log_{10} \frac{3}{2} + 10 \log_{10} (2^{2l})$$

or $\quad \dfrac{S}{N_Q} = 1 \cdot 76 + 6 \cdot 02 \, l \text{ dB}$

Other definitions may be given, but these generally differ only by a constant when the number of levels is large. Some values are given in the table below. Typically an eight-binit word is found adequate for a speech link, while 10–12 binits would be required for music transmissions of high quality. The more binits used to specify the sample voltage, the better the signal-to-quantisation-noise ratio but at the expense of a wider bandwidth for the signal during transmission. Once again we meet the idea of an exchange rate between these quantities (Table 7.2).

TABLE 7.2

L	l	S/N_Q (dB)
16	4	25·8
32	5	31·9
64	6	37·9
128	7	43·9
256	8	49·9
512	9	55·9
1024	10	62·0
2048	11	68·0
4096	12	74·0

C. Detecting Signals of Known Form

7.12 BINARY COMMUNICATION SYSTEMS

Pulse code modulation is one example of communication by binary signalling; the more familiar teleprinter signal is another. In these signals two different waveforms are employed, one to represent

1 and a second to represent 0. The waveforms may be 'pulse' and 'no pulse', as we assumed in the previous section, or else two different wave shapes having the same energy may be employed, e.g. positive and negative pulses. We may assume that synchronisation can be established so that the essentials of the problem to be discussed are as follows: In a given time interval, a noisy waveform is received. A signal of known shape may or may not be present. Is there a 'best' way of deciding whether or not this signal is present? A variant of the problem asks which of two different signals, each of known form, is present.

To introduce the concepts involved in tackling this problem let us begin with the simplest signal waveform possible—a steady d.c. signal. The problem is by no means trivial for it is the problem faced by the experimentalist who wishes to detect and measure a small steady voltage or current.

7.13 D.C. SIGNALS. INSTANTANEOUS OBSERVATION

Suppose an observation of the received waveform $v(t)$ is made at time t and that the noise is Gaussian with mean power (mean square voltage) σ^2. The observed voltage may or may not contain a contribution due to a d.c. signal component, s. If the signal is not present, then the probability that v lies in the range v, $v+\mathrm{d}v$ is given by

$$P(v \mid \text{not } s) = \frac{1}{\sqrt{(2\pi\sigma^2)}} \, e^{-v^2/2\sigma^2}$$

while if the signal is present

$$P(v \mid s) = \frac{1}{\sqrt{(2\pi\sigma^2)}} \, e^{-(v-s)^2/2\sigma^2}$$

Note that these are statements of conditional probability (see Chapter 5). But our problem asks for the inverse probability: given v, what is the chance that a signal of strength s is present? We may attempt to calculate this using the rule relating conditional probabilities (Bayes's formulae) given in Section 5.3

$$P(s \mid v) = \frac{P(s)\,P(v \mid s)}{P(s)\,P(v \mid s) + P(\text{not } s)\,P(v \mid \text{not } s)}$$

$$P(s \mid v) = \frac{\Lambda P(s)/P(\text{not } s)}{1 + \Lambda P(s)/P(\text{not } s)} \qquad 7.36$$

where

$$\Lambda = \frac{P(v|s)}{P(v|\text{not } s)} = e^{-s(s-2v)/2\sigma^2}$$

If v is large and positive, then Λ is large and equation 7.36 gives almost unit probability for the presence of the signal. If v is large and negative then Λ is small and so is $P(s/v)$. This, of course, is in agreement with personal judgement.

The ratio expressed by Λ is termed a *likelihood ratio* since a conditional probability $P(Y|X)$, considered as a function of the hypothesis X, is designated a *likelihood function* in the theory of statistical inference.

Unfortunately, equation 7.36 does not provide a complete answer because the observer cannot in general be expected to know $P(s)$ and $P(\text{not } s)$ which are termed the *a priori probabilities*. One way out of this dilemma is to equate our lack of knowledge as to whether or not the signal exists by writing

$$P(s) = P(\text{not } s) = 1/2$$

This approach, first suggested by Bayes in a different context, brings equation 7.36 to the form

$$P(s/v) = \frac{\Lambda}{1+\Lambda}$$

On this basis, the signal has a 50% chance of being present when $\Lambda = 1$. In our present example, this corresponds to $v = s/2$. This, of course, is precisely where we would have set the decision threshold in the first place but it is now clear that the guesswork involved relates to an estimate of the *a priori* probabilities. If $P(s) > P(\text{not } s)$ then equation 7.36 would set a decision threshold somewhat lower and at a level which may be calculated if these *a priori* probabilities are known and assuming that we desire a 50% chance of making a correct decision at threshold. (This last assumption implies that we consider that the 'cost' of making an error either way is the same.) Hypothesis testing, as this procedure is called, forms a major part of statistical communication theory.

7.14 D.C. SIGNALS—OBSERVATION OVER AN INTERVAL T

So far we have permitted the signal to take only one of two possible values, s or 0. The problem may be changed by allowing the signal voltage to have any value with equal probability within an unspecified but very wide range. What is the best way of estimating the value, s, of an *unknown* d.c. signal? (See also Woodward.[4])

The probabilities $p(s)$ and $p(s|v)$ now become probability density functions, thus

$$p(s|v) = \frac{p(s)\,p(v|s)}{\int p(s)\,p(v|s)\,\mathrm{d}s} \qquad 7.37$$

Because we are assuming $p(s)$ to be a constant, this factor cancels. Moreover

$$\int p(v|s)\,\mathrm{d}s = \int \frac{1}{\sqrt{(2\pi\sigma^2)}}\,\mathrm{e}^{-(v-s)^2/2\sigma^2}\,\mathrm{d}s$$

so that integrating over a sufficiently wide range this integral is unity, hence

$$p(s|v_1) = p(v_1|s)$$

$$= \frac{1}{\sqrt{(2\pi\sigma^2)}}\,\mathrm{e}^{-(v_1-s)^2/2\sigma^2} \qquad 7.38$$

where the notation v_1 is introduced to emphasise that this is the result of a single observation at $t = t_1$. Equation 7.38 gives the probability that the d.c. signal has the value s on the basis of a single measurement, v_1. Suppose now that a second, independent observation, v_2, is made. The new conditional probability density $p(s|v_1, v_2)$ may be found using equation 7.37 but now, as we have some knowledge of s from the first measurement (expressed by equation 7.38), the function $p(s/v_1)$ replaces the *a priori* probability $p(s)$. Hence

$$p(s|v_1, v_2) = \frac{1/(2\pi\sigma^2)\,[\exp-(v_1-s)^2/2\sigma^2]\,[\exp-(v_2-s)^2/2\sigma^2]}{1/2\pi\sigma^2 \int [\exp-(v_1-s)^2/2\sigma^2]\,[\exp-(v_2-s)^2/2\sigma^2]\,\mathrm{d}s}$$

This is clearly normalised with respect to s, as must be the case.

234

We shall have no further interest in the normalising constant so we write

$$p(s|v_1, v_2) = \text{const.} \times \exp\{-[(v_1-s)^2+(v_2-s)^2]/2\sigma^2\}$$

If N successive independent observations are made

$$p(s|v_1, v_2, v_3 \ldots v_N) = \text{const.} \times \exp\left\{-\sum_{n=1}^{N}(v_n-s)^2/2\sigma^2\right\}$$

But N independent observations are equivalent to sampling at intervals π/W for a time $T = N\pi/W$. The summation may be replaced by an integral over time, using the results of Section 2.16 (accurate only for large N)

$$p(s|v(t)) = \text{const.} \times \exp-\frac{W}{2\pi\sigma^2}\int_0^T [v(t)-s]^2\, \mathrm{d}t \qquad 7.39$$

The most probable value, s_m, is found by minimising the integral by varying s. Differentiating under the integral sign and equating to zero yields

$$s_m = \frac{1}{T}\int_0^T v(t)\, \mathrm{d}t \qquad 7.40$$

so the most probable value for s is equal to the mean value of the signal. This comes as no surprise, but it is pleasing to be able to obtain this result from a statistical argument.

Three important conclusions may be drawn from this analysis. First, equation 7.40 suggests that an optimum way to process a noisy d.c. signal is to integrate (or multiply by a constant and integrate) the received waveform over the period T in which it is known to exist. We shall see later that this corresponds to *correlation detection*.

Second, the operation expressed by equation 7.40 would be performed electrically by an averaging filter of the kind discussed in the analysis of shot noise (Section 6.4). This is essentially a low-pass filter whose response, at time $t = T$, is s_m. In the context of the material to follow, this operation may be described as optimum or *matched filter* detection. In passing we note that s_m is identical to the d.c. Fourier component of the waveform when expanded in the range $t = 0$ to $t = T$. This may be regarded as a random variable, whose variance was shown in Section 7.4 (equation 7.12) to be σ^2/N. The square root of this quantity gives the standard deviation, or standard error of the measurement

$$\sqrt{\left(\frac{\sigma^2}{N}\right)} = \sqrt{\left(\frac{\pi\sigma^2}{WT}\right)} = \sqrt{\left(\frac{\pi p_n}{T}\right)}$$

which reduces as the square root of the observation time. To measure a noisy d.c. signal one would select an instrument having a long time constant, or narrow bandwidth.

Finally, we note that minimising the integral in equation 7.39 is equivalent to minimising the mean square difference between $v(t)$ and s. It is essentially a *least squares fitting* procedure.

7.15 MATCHED-FILTER: SIGNAL-TO-NOISE RATIO APPROACH

In the previous section we were presented with the problem of detecting a simple signal of known form, namely d.c., which was known to be present in a time interval 0–T. One suitable method of processing was to pass the signal through a linear filter having an impulse response

$$h(\tau) = 1 \qquad 0 < \tau < T$$
$$h(\tau) = 0 \qquad |\tau| > T$$

and the output of such a filter at time T is precisely the integral of equation 7.40.

Generalising this problem, suppose that the signal, if present in the interval T, is known to have the form $s(t)$. Is there a 'best' filter to use in this case? It turns out that there is and that it is best specified in terms of its impulse response. We will take a signal-to-noise ratio approach and first imagine the signal to be passed through a linear filter of arbitrary impulse response $h(\tau)$ *in the absence of noise* and compare the square of this output voltage at the moment $t = T$ with the mean square output voltage when *white Gaussian noise alone* is presented at the input. As the filter is linear, the result will be compatible with the other definition of S/N given in Section 7.6. Calling the signal response $g_s(T)$, we have

$$g_s(T) = \int_{-\infty}^{\infty} s(\tau)\, h(T-\tau)\, \mathrm{d}\tau = \int_{-\infty}^{\infty} s(T-\tau)\, h(\tau)\, \mathrm{d}\tau$$

$$g_s^2(T) = \left[\int_{-\infty}^{\infty} s(T-\tau)\, h(\tau)\, \mathrm{d}\tau \right]^2$$

7.41

The range of integration may be restricted to $0 < \tau < T$ since, for the present, we assume $s(\tau)$ to be zero outside this range. With white noise of power spectral density (one-sided) p_n, the output noise voltage has a mean power (volts squared)

$$\overline{g_N^2(T)} = \int_{-\infty}^{\infty} \frac{p_n |H(\omega)|^2}{2} \, d\omega$$

where the factor of two allows for the inclusion of the negative half of the frequency axis. Transforming this last equation by Parseval's theorem (equation 2.45)

$$\overline{g_N^2(T)} = \pi p_n \int_{-\infty}^{\infty} h^2(\tau) \, d\tau$$

Thus the signal-to-noise ratio may be written

$$\frac{g_s^2(T)}{\overline{g_N^2(T)}} = \frac{1}{\pi p_n} \frac{\left[\int s(T-\tau) h(\tau) \, d\tau \right]^2}{\int h^2(\tau) \, d\tau}$$

multiplying above and below by the signal energy E, where

$$E = \int_{-\infty}^{\infty} s^2(\tau) \, d\tau = \int_{-\infty}^{\infty} s^2(T-\tau) \, d\tau$$

$$\frac{g_s^2(T)}{\overline{g_N^2(T)}} = \frac{E}{\pi p_n} \frac{\left[\int s(T-\tau) h(\tau) \, d\tau \right]^2}{\left[\int s^2(T-\tau) \, d\tau \right] \times \left[\int h^2(\tau) \, d\tau \right]} \qquad 7.42$$

This quantity has to be maximised by varying the function $h(\tau)$. Happily there is a mathematical theorem, known as Schwarz's inequality, which gives the result directly. This states that for two functions $f_1(x)$ and $f_2(x)$

$$\left[\int_a^b f_1(x) . f_2(x) \, dx \right]^2 \leqslant \left[\int_a^b f_1^2(x) \, dx \right] \times \left[\int_a^b f_2^2(x) \, dx \right]$$

The equality holds if

$$f_2(x) = \alpha f_1(x)$$

where α is a (positive or negative) constant.

237

It follows that the expression on the right-hand side of equation 7.41 takes a maximum value of

$$\left[\frac{g_s^2(T)}{\overline{g_N^2(T)}} \right]_{max} = \frac{E}{\pi p_n} \qquad 7.43$$

when

$$h(\tau) = \alpha s(T-\tau) \qquad 7.44$$

Note that the maximum possible signal-to-noise ratio depends only on the total energy of the signal and the noise power density and not at all on the shape of the signal waveform. Indeed any other result would have been surprising, for with white Gaussian noise it can make no difference how the signal energy is distributed along either the time or frequency axes. In particular, for a d.c. signal, s,

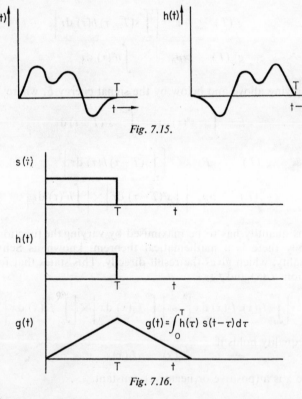

Fig. 7.15.

$$g(t) = \int_0^T h(\tau) \, s(t-\tau) \, d\tau$$

Fig. 7.16.

we have $E = s^2 T$ and hence, in this case and with the notation of Section 7.4, the signal-to-noise ratio becomes s^2/a_0^2 as expected.

A filter having the impulse response given by equation 7.44 is termed a *matched filter*, the 'match' being to the shape of the expected signal waveform. The impulse response desired has the shape of the signal waveform run backwards from the moment $t = T$. Fig. 7.15 illustrates this result. In Fig. 7.16 the response from a filter matched to a rectangular pulse of duration T is shown. A decision regarding the existence or absence of the signal would be based on

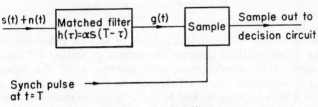

Fig. 7.17. Matched filtering

the observed output voltage at time $t = T$ (Fig. 7.17). Our concern here has been to deduce the best signal processing method possible from a theoretical standpoint. The physical realisibility of a matched filter is a separate question.

7.16 MATCHED-FILTER EQUIVALENT TO CORRELATION DETECTION

A matched filter with the impulse response given by equation 7.44 will have as its output at time T the voltage given by equation 7.41, namely

$$g_s(T) = \int_0^T s(\tau) \times [\alpha s(\tau)] \, d\tau$$

$$= \alpha \int_0^T s^2(t) \, dt$$

Thus a result identical to matched filtering would be obtained by multiplying the incoming signal with a synchronous noise-free signal generated in the receiver and integrating the result over the duration of the signal. This procedure, known as correlation detection, is illustrated in Fig. 7.18.

239

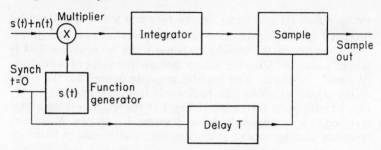

Fig. 7.18. Correlation detection

The detection of sinusoidal signals of known phase by means of the 'phase-sensitive detector' mentioned in Section 7.7 uses this principle. After amplification to a suitable level, the incoming signal is multiplied, using a balanced modulator circuit, by a sinusoid which is phase-locked to the expected signal. The low-frequency difference terms are passed on to integrating circuits which effectively reduce the observation bandwidth to a value $1/4CR$ Hz which can be made very narrow, provided long response times can be tolerated. (See Chapter 6, Problem 1.)

7.17 SYNCHRONOUS BINARY SIGNALLING: LEAST SQUARES FITTING APPROACH

Consider a binary signalling system in which the signals transmitted are either $s_1(t)$ or $s_2(t)$ representing 'mark' and 'space', respectively. The waveform $f(t)$ at the receiver contains additive noise $n(t)$ so that

either $\qquad\qquad f(t) = s_1(t) + n(t)$

or $\qquad\qquad f(t) = s_2(t) + n(t)$

Synchronising pulses are assumed to be available at the receiver at the moments dividing the intervals T, occupied either by 'mark' or 'space'. The problem at the receiver is then to process $f(t)$ in some way which will enable a decision to be made as to whether $s_1(t)$ or $s_2(t)$ was transmitted. Because of the noise, the probability of making a correct decision is less than unity and the object of the ideal receiver is to make this probability a maximum. Equivalent ways in which this might be accomplished have been discussed in the preceding two sections. For variety we take a third approach which leads to the same result.

Let the received waveform in any given interval T be compared with two locally generated waveforms $ks_1(t)$ and $ks_2(t)$ and a decision be based on finding which gives the best least squares fit to the received waveform. That is to say, we decide that 'mark' was sent if

$$\int_0^T [f(t) - ks_1(t)]^2 \, dt < \int_0^T [f(t) - ks_2(t)]^2 \, dt$$

and that 'space' was sent otherwise.

For simplicity assume that $s_1(t)$ and $s_2(t)$ have equal energy

$$\int_0^T s_1^2(t) \, dt = \int_0^T s_2^2(t) \, dt = t$$

The inequality above now reads

$$\int_0^T f(t) \, s_1(t) \, dt < \int_0^T f(t) \, s_2(t) \, dt$$

but these are the two outputs of correlation detectors or matched filters designed for $s_1(t)$ and $s_2(t)$. These have already been shown to give optimum signal-to-noise performance. A receiver which would base its decision on the inequality above, using matched filters, is shown in Fig. 7.19. A correlation method would perform identically.

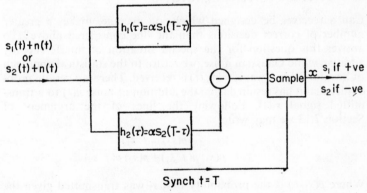

Fig. 7.19. Optimum receiver

241

So far, nothing has been said about the shapes of $s_1(t)$ and $s_2(t)$; we have only stipulated equal energy. Suppose that $s_1(t)$ was sent. In the absence of noise our receiver inspects the difference

$$\Delta = \int_0^T s_1^2(t) \, \mathrm{d}t - \int_0^T s_1(t) \, s_2(t) \, \mathrm{d}t \qquad 7.45$$

and it is clearly desirable for this quantity to be as large as possible. The first term is the signal energy E. Since this is held constant it follows from Schwartz's inequality that Δ will be a maximum for $s_2(t) = -s_1(t)$, when $\Delta = 2E$. For example, $s_1(t) = \cos \omega t$ and $s_2(t) = \cos(\omega t + \pi)$ would give this result and the term 'coherent phase reversal keying' is applied to a binary signalling system with these alternatives for 'mark' and 'space'. Easier to implement is 'frequency shift keying' in which two carriers of different frequencies are used for $s_1(t)$ and $s_2(t)$, with random phase relationship. The cross-correlation term in equation 7.45 is then zero, i.e. these are orthogonal signals, and $\Delta = E$. This system would thus be 3 dB poorer than the optimum. Apart from considerations directed towards maximising Δ, however, the choice of wave-shapes for $s_1(t)$ and $s_2(t)$ is arbitrary as far as signal-to-noise ratio is concerned. As one writer (R. M. Lerner in Baghdady[5]) has remarked, there is no signal which is 'least like' Gaussian noise. It is as though white Gaussian noise, like the devil of a different hue, may wear any disguise in order to deceive.

7.18 STATISTICAL DECISIONS

Can a receiver be designed which, on average, makes a greater number of correct decisions than the one considered above? To answer this question for the special situation of interference by additive white Gaussian noise, we return to the statistical approach of Section 7.13. A waveform $f(t)$ is received. There is a certain probability that it has resulted from the addition of noise $n(t)$ to a transmitted signal $s_1(t)$. Following the lines of the argument of Section 7.13 we may write

$$p(s_1 | f) = \frac{p(s_1) \, p(f | s_1)}{p(s_1) \, p(f | s_1) + p(s_2) \, p(f | s_2)}$$

Where $p(s_1 | f)$ is the probability that s_1 was transmitted given the observed waveform f. Similarly, the probability that s_2 was transmitted is

$$p(s_2 \mid f) = \frac{p(s_2)\,p(f \mid s_2)}{p(s_1)\,p(f \mid s_1) + p(s_2)\,p(f \mid s_2)}$$

Assuming, as before, equal *a priori* probabilities for s_1 and s_2

$$p(s_1 \mid f) = \frac{\Lambda}{1 + \Lambda}$$

and

$$p(s_2 \mid f) = \frac{1}{1 + \Lambda} \qquad\qquad 7.46$$

where

$$\Lambda = \frac{p(f \mid s_1)}{p(f \mid s_2)} \qquad\qquad 7.47$$

is the likelihood ratio.

A statistical decision as to whether s_1 or s_2 was transmitted must be based on equations 7.46. If $\Lambda > 1$ then it is more probable that s_1 was transmitted; if $\Lambda < 1$ then the chances are that s_2 was sent. Clearly, no receiver can be constructed which will, on average, make a greater number of correct decisions than one which uses the statistical decision criterion, 'compare Λ with unity'. To see how this criterion compares with the electrical processing of the signal just described we must first compute the likelihood ratio.

The probability $p(f \mid s_1)$ is the probability that the noise has changed the signal from $s_1(t)$ to $f(t)$ which means that at each sampling instant t_k

$$n(t_k) = f(t_k) - s(t_k)$$

The $n(t_k)$ at each sampling instant are independent with a Gaussian distribution of variance σ^2 and hence the product law for independent events gives

$$p(f \mid s_1) = \frac{1}{\sqrt{(2\pi\sigma^2)}^{WT/\pi}} \exp - \sum_k \frac{[f(t_k) - s_1(t_k)]^2}{2\sigma^2}$$

Hence for the likelihood ratio (equation 7.47)

$$\log \Lambda = - \sum_k \frac{[f(t_k) - s_1(t_k)]^2 - [f(t_k) - s_2(t_k)]^2}{2\sigma^2}$$

If W, the bandwidth within which the signals and noise are received, is sufficiently large then as before (Section 7.14) the summation may be replaced by an integral over time

$$\log \varLambda = -\frac{W}{2\pi\sigma^2} \left[\int_0^T [f(t) - s_1(t)]^2 \, \mathrm{d}t - \int_0^T [f(t) - s_2(t)]^2 \, \mathrm{d}t \right]$$

When, as we assumed in the previous section, the energies of the signals s_1 and s_2 are the same, this expression simplifies and the decision rule, 'compare $\log \varLambda$ with 0', becomes 'compare

$$\left\{ \int_0^T f(t) \, s_1(t) \, \mathrm{d}t - \int_0^T f(t) \, s_2(t) \, \mathrm{d}t \right\} \quad \text{with } 0'.$$

This is precisely the operation which the receiver considered in the previous section performed. With the assumptions made here, namely $p(s_1) = p(s_2)$, $T \gg 1/W$, white Gaussian noise and equal energy signals, this detection system may truly be regarded as optimum.

7.19 ERROR RATES IN BINARY SIGNALLING

The probability that the optimum receiver of Fig. 7.19 will make an incorrect decision as to whether s_1 or s_2 was transmitted may be found as follows. It is assumed that $s_1(t)$ and $s_2(t)$ carry equal energy E and that they are orthogonal, thus

$$\int_0^T s_1^2(t) \, \mathrm{d}t = \int_0^T s_2^2(t) \, \mathrm{d}t = E$$

and

$$\int_0^T s_1(t) \, s_2(t) \, \mathrm{d}t = 0$$

Now suppose that $s_1(t)$ is received in the absence of noise. At the sampling instant $t = T$ the voltages at A and B are given by

$$v_A = \int_0^T s_1(\tau) \, h_1(T-\tau) \, \mathrm{d}\tau$$

$$= \alpha \int_0^T s_1^2(\tau) \, \mathrm{d}\tau$$

$$= \alpha E$$

$$v_B = \int_0^T s_1(\tau) \, h_2(T-\tau) \, \mathrm{d}\tau = 0$$

The sample voltage, x_0, passed on to the decision circuit in the noiseless situation is therefore

$$x_0 = v_A - v_B = \alpha E$$

When noise is present, the sample x will be a value of a Gaussian random variable whose mean is x_0 and whose variance σ^2 is the sum of the noise powers $\overline{n_A^2}$ and $\overline{n_B^2}$ passed by the matched filters. If p_n is the one-sided input noise power density in v^2 rad^{-1} s (defined in Section 7.3) then

$$\overline{n_A^2} = p_n \int_0^\infty |H_1(\omega)|^2 \, d\omega$$

$$= \pi p_n \int_0^T h_1^2(t) \, dt$$

$$= \pi \alpha^2 p_n E$$

where equation 2.45 has been used to transform the integral from the frequency domain to the time domain, recalling that $h_1(t)$ is zero outside the range of integration shown. An identical expression is obtained for $\overline{n_B^2}$ hence

$$\sigma^2 = \overline{n_A^2} + \overline{n_B^2} = 2\pi\alpha^2 p_n E$$

The probability function for the sample voltage x is therefore (equation 5.32)

$$p(x) = \frac{1}{\sqrt{(2\pi\sigma^2)}} \, e^{-(x-\alpha E)^2/2\sigma^2}$$

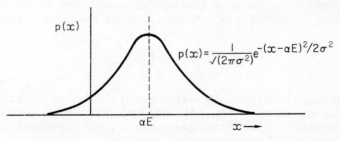

$$p(x) = \frac{1}{\sqrt{(2\pi\sigma^2)}} e^{-(x-\alpha E)^2/2\sigma^2}$$

Fig. 7.20.

245

which is sketched in Fig. 7.20. The decision 's_1' is made whenever $x > 0$; the probability w of a wrong decision is therefore given by

$$w = \int_{-\infty}^{0} p(x)\, dx$$

$$= \frac{1}{\sqrt{(2\pi\sigma^2)}} \int_{\alpha E}^{\infty} e^{-u^2/2\sigma^2}\, du \quad (u = \alpha E - x)$$

$$= \frac{1}{\sqrt{(2\pi\sigma^2)}} \int_{0}^{\infty} e^{-u^2/2\sigma^2}\, du - \frac{1}{\sqrt{(2\pi\sigma^2)}} \int_{0}^{\alpha E} e^{-u^2/2\sigma^2}\, du$$

$$= \frac{1}{2} \left[1 - \sqrt{\frac{2}{\pi\sigma^2}} \int_{0}^{\alpha E} e^{-u^2/2\sigma^2}\, du \right]$$

$$= \frac{1}{2} \left[1 - \mathrm{erf}\left(\frac{\alpha E}{\sqrt{(2\sigma^2)}} \right) \right]$$

$$= \frac{1}{2} \left[1 - \mathrm{erf}\left(\sqrt{\frac{E}{4\pi p_n}} \right) \right] \qquad 7.48$$

By symmetry, the same error probability applies for mistaking s_2 for s_1. In this equation E is the total signal energy measured in V^2 s (or A^2 s) and $2\pi p_n$ is the mean square input noise voltage received in each 1 Hz bandwidth.

The assumption that s_1 and s_2 are orthogonal may be made valid for sinusoids of different frequency, so that equation 7.48 gives the minimum error probability possible with frequency-shift keying (Fig. 1.5(c)). As noted in Section 7.17, a better performance is possible when $s_1(t) = -s_2(t)$ which is the case for phase-shift keying (Fig. 1.5(d)) with 180° phase shifts. The analysis for this system follows along the same lines as that given above, except that $v_B = -\alpha E$ instead of $v_B = 0$. Thus $x_0 = 2\alpha E$ and the final result is

$$w = \frac{1}{2} \left[1 - \mathrm{erf}\left(\sqrt{\frac{E}{2\pi p_n}} \right) \right] \qquad 7.49$$

showing a 3 dB superiority over frequency shift keying. These error rates are plotted in Fig. 7.21; they have been derived assuming matched-filter detection but the same result would be obtained using a correlation detector method.

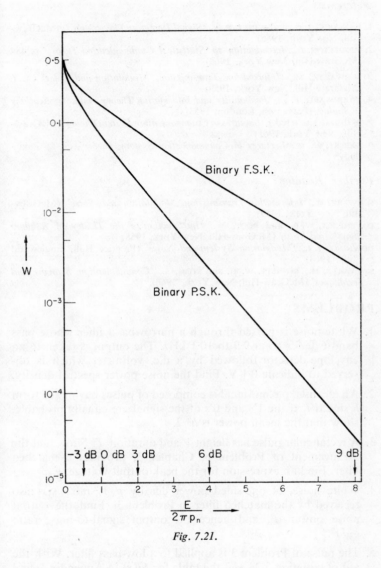

Fig. 7.21.

References

1. HANCOCK, J. C., and WINTZ, P. A., *'Signal Detection Theory'* ch. 2 (McGraw-Hill, New York, 1966)
2. MIDDLETON, D., *Introduction to 'Statistical Communication Theory'* p. 484 (McGraw-Hill, New York, 1960)
3. SCHWARTZ, M., *'Information Transmission, Modulation and Noise'* ch. 6 (McGraw-Hill, New York, 1959)
4. WOODWARD, P. M., *'Probability and Information Theory, with Applications to Radar'* (Pergamon, London, 1953)
5. BAGHDADY, E. J. (ed.), *'Lectures in Communicatiion System Theory'* (McGraw-Hill, New York, 1961)
6. SIEBERT, W. M., *'Lectures in Communication System Theory'*, ch. 8, Baghdady[5]

Further Reading

SCHWARZ, M., *'Information Transmission, Modulation and Noise'* (McGraw-Hill, New York, 1959)
DAVENPORT, W. B., and ROOT, W., *'Introduction to the Theory of Random Signals and Noise'* (McGraw-Hill, New York, 1958)
DOWNING, J. J., *'Modulation System and Noise'* (Prentice Hall, Englewood Cliffs, 1964)
SCHWARTZ, M., BENNETT, W. R., and STEIN, S., *'Communication Systems and Techniques'* (McGraw-Hill, New York, 1966).

PROBLEMS

1. White noise is passed through a narrowband filter whose pass band extends from 9·9 to 10·1 kHz. The output is taken to an envelope detector followed by a d.c. voltmeter which is observed to indicate 0·1 V. Find the noise power spectral density.

2. An idealised p.c.m. signal is composed of pulses having the form $h \, \text{sinc} \, Wt$. If the 1's and 0's of the signal are equally probable, show that the mean power is $h^2/2$.

3. A rectangular pulse has height V and duration T. Show that the arrangement of Problem 6, Chapter 3, provides a matched filter. Find an expression for the peak output voltage, g_m.

4. White noise, of one-sided power density $p_n \, V^2 \, \text{rad}^{-1} \, \text{s}$ is also received by the matched filter in Problem 3. Find the output noise power σ^2, and hence the output signal-to-noise ratio g_m^2/σ^2.

5. The pulse of Problem 3 is applied to a low-pass filter. With the aid of equation 3.24 and the table for $Si(x)$ in Appendix 1, find approximately the cut-off frequency for maximum signal-to-

noise ratio at the output and compare this signal-to-noise ratio with the result of Problem 4.

6. Two noise signals of duration T s and bandwidth W rad s^{-1} are given by (cf. equation 7.2)

$$a(t) = \sum_k a_k \operatorname{sinc} W\left(t - \frac{k\pi}{W}\right)$$
$$b(t) = \sum_k b_k \operatorname{sinc} W\left(t - \frac{k\pi}{W}\right) \qquad -\frac{WT}{2\pi} < k < \frac{WT}{2\pi}$$

Show that their inner product (Appendix 3) is

$$\mathbf{A} \cdot \mathbf{B} = \sum_k a_k b_k$$

Show also, by considering the mean and variance of $\mathbf{A} \cdot \mathbf{B}$, that the signals are almost certainly orthogonal. (Geometrically this means that two vectors chosen at random, in a space of many dimensions, are almost certain to lie at right angles.)

7. A binary coded signal is received from a distant space station against sky noise at 29°K. The signal is modulated by phase reversal keying at 100 binary symbols per second and the received signal power is 10^{-19} W. Find, from Fig. 7.21, the minimum possible error rate.

Chapter 8

Information and Coding

8.1 INTRODUCTION

We have so far been almost exclusively concerned with studying the structure and properties of communication *signals*. A signal waveform may be interesting in its own right, it may have an unusual spectrum, it may have been modulated, filtered, sampled, or cross-correlated in some highly ingenious and noise-baffling way. The signal is, nevertheless, only a means to an end, for the function of a communication system is to transmit not signals, but *information* from place to place. We may think of a communication link as handling the commodity 'information', using the signal as a messenger. The success of the system in transmitting signals, the mean received power, percentage distortion, etc., may be assessed in terms of conventional electrical measurements, but how about its success as a conveyor of information? Can information be measured?

In everyday usage the word 'information' is associated in our minds with 'knowledge', 'given facts' or 'news', and dictionary definitions generally list these as near synonyms. In communication theory, however, this term is given a much more restricted definition, so much so that it might have been better to have invented a new word in order to avoid confusion. Yet when we encounter a sentence such as, 'Many were persuaded by the force of his argument', we do not think to measure the 'force' in newtons and a similar semantic filter must be used when the term 'information' is encountered in non-scientific contexts. The important thing about information in its engineering sense is that it may be *measured*, and a body of theory has been developed which enables the engineer to calculate the fastest rate of information transmission that is possible for a given link, under known conditions of bandwidth and signal-to-noise ratio.

Information theory is not so much concerned with how to design a particular communication system as with the much broader concepts of what is intrinsically possible and what is certainly impossible under given conditions, and with providing sound criteria by which different systems may be compared. The relationship between information theory and communication systems has been likened to that between thermodynamics and heat engines. The second law of thermodynamics states what is possible and what is not possible as regards heat engines and leads to a proper measure of thermodynamic efficiency. Similarly, the coding theorems of information theory, while not providing very practical design techniques for ideal codes, enable the designer to assess the efficiency of his system or prevent him from attempting the impossible.

It was the existence of heat engines which stimulated the development of thermodynamics. In a similar way, information theory did not appear as a distinct discipline until communication systems were widely established. Some of the fundamental ideas stem from the work of Hartley and of Nyquist in the 1920s, and other roots go back much earlier, but information theory received its major impetus from the work of Shannon, whose famous paper 'A mathematical theory of communication' was published in 1948. In these final chapters, some of the elementary principles of the theory will be outlined, in the hope that this will whet the reader's appetite and stimulate him to explore for himself the large volume of literature now devoted to this field.

8.2 MEASURING INFORMATION

Suppose a stranger to a certain building sees three light switches on a wall and enquires which one turns on the first-floor lights. He then expects to receive one of the three messages, 'the left-hand one', 'the middle one', or 'the right-hand one', each of these messages being equally likely as far as he is concerned. On being told 'the right-hand one' he has received a certain amount of information. Now suppose that there had been nine switches, in three rows of three. His initial uncertainty would have been greater and the message 'the right-hand one on the top row' carries more information than in the simpler case of only three switches. The greater the uncertainty as to which message will arrive, the greater the information content of the message. Information gain is equivalent to removal of uncertainty.

The situation in this example is typical of any in which 'information' is transferred, inasmuch as the recipient is expecting one of a number of possible messages. When we receive a telegram we receive one of a large but finite number of intelligible word combinations which are of convenient length for a telegram. When listening to a weather forecast we expect to be told of one of a number of possible weather conditions. Given that the message is one of N possible alternative messages, the amount of information conveyed may be measured as some monotonically increasing function of N. We assume for the moment that all N messages are equally probable.

Returning to the light switch problem, we may say that the specification of one of three possibilities, 'left', 'middle', or 'right', resolves one three-fold choice; when there are nine switches, two three-fold choices are involved. Had there been three panels of nine switches, three three-fold instructions would have been required in order to specify one out of 27 switches. The tree diagram of Fig. 8.1 shows how the number of alternative 'messages' grows exponentially with the number of three-fold instructions needed to specify a particular message. Yet, intuitively, we feel that each of the instructions carries the same amount of information—since each resolves a three-fold choice. Taking, therefore, the number of instructions required to specify a message as a measure of the information it carries, we have the following conclusion: *the amount of information conveyed in specifying one of a number of possible and equally likely alternatives (messages) is measured as the logarithm of the number of alternatives.*

$$I = \log N \qquad\qquad 8.1$$

Thus, taking logarithms to base 3, we have $I = \log_3 3 = 1$ three-fold unit of information required when there are three light switches, $I = \log_3 9 = 2$ three-fold units for nine switches, etc.

The base of logarithms determines the unit. A tree diagram like Fig. 8.1, but using two-fold choices, could be drawn showing how $\log_2 N$ two-fold units of information are sufficient to identify one of N messages. A binary communication system in fact transmits

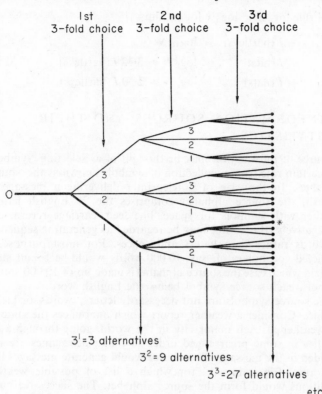

1st 3-fold choice	2nd 3-fold choice	3rd 3-fold choice

$3^1 = 3$ alternatives

$3^2 = 9$ alternatives

$3^3 = 27$ alternatives

etc.

Fig. 8.1. Showing how three three-fold instructions, namely '133', are sufficient to lead from 0 to the point starred, i.e. to specify one of 27 messages

information as a sequence of two-fold choices. If N messages were listed in a code book, each message being numbered using the binary system, then any message may be transmitted over a binary channel simply by sending its binary number and this would require the transmission of $\log_2 N$ binary digits (including any initial zeros). The 'two-fold' unit of information is called the *bit* (abbreviated from 'binary digit'). The idea of a logarithmic measure of information is due to Hartley (1924) and when common (base 10) logarithms are used, the unit is the *hartley*. In mathematical analysis natural logarithms are more appropriate and the unit is then, unhappily, termed the *nat*. Conversion among the units follows the

253

usual rule $\log_a x = \log_b x/\log_b a$, thus

$$I \text{ (hartleys)} = \log_{10} N$$
$$I \text{ (bits)} = \log_2 N = 3 \cdot 32 \, I \text{ (hartleys)}$$
$$I \text{ (nats)} = \log_e N = 2 \cdot 30 \, I \text{ (hartleys)}$$

8.3 INFORMATION SOURCES AND THEIR STATISTICS

A source of information may be thought of as selecting 'symbols' at a certain rate from a collection of symbols known as the 'source alphabet'. In the case of a teleprinter delivering a message in English, the source alphabet comprises the 26 English letters, together with symbols for space, line feed, carriage return, etc. Alternatively, this source may be regarded as generating sequences of words rather than letters and spaces. For most purposes a restricted vocabulary of, say, 10 000 words would be found sufficient, in which case the source alphabet is made up of 10 000 source symbols, each source symbol being one English word.

The source symbols are not necessarily letters, words or hieroglyphics. Consider a weather report which announces the state of the weather at each major city in the world, going through a list of cities in some prearranged order so that place names are not included in the message. This source would generate messages like 'sun, rain, fog, rain ...' for which a list of possible weather conditions would form the source alphabet. The successive pulse heights of a quantised p.a.m. signal form a series of numbers (voltages) selected from a finite source alphabet which includes all the available quantisation levels. Any source which transmits information piecemeal, item by item, may be viewed in this way, the list of alternative items forming the source alphabet.

An important property of an information source is the *source entropy* which is a measure of the average amount of information transmitted per source symbol. In the simplest case of an alphabet of N symbols, each occurring with equal probability, the argument of the preceding section would set the source entropy at $\log_2 N$ bits per symbol. If English text, with an alphabet of 26 letters $+1$ space, behaved in this way then $\log_2 27 = 4 \cdot 75$ bits of information would be transmitted for each letter selected. But English text is a much more complex information source then this. First, the letters do not appear with equal relative frequencies (for example, in a long section of text about 0·12 of the letters will be E, while only 0·02 will be w. Secondly, the probability of the source selecting a given letter is not a unique quantity but depends on the preceding letters. For example, if the letter Q occurs, the probability is almost unity that the next letter will be U, so that when the letter U appears it carries very little information with it, because there was little doubt that it would in fact turn up as the next letter. A letter U preceded by any letter other than Q carries much more information.

To take account of the second complication is beyond the scope of this discussion and we will limit ourselves to information sources whose symbols occur with mutually independent probabilities. These are sometimes called 'zero memory' sources, the likelihood of occurrence of any symbol being unchanged by the preceding symbol or symbols. The restriction to zero memory sources is a severe one and will make the examples chosen to illustrate the theory somewhat artificial, although profitable results can still be obtained. The complication of unequal probabilities for the occurrence of the source symbols is less serious and may be tackled as follows. In the case when all N symbols have equal probability, $p = 1/N$, the information per symbol is, from the argument already presented

$$I = \log N = \log 1/p$$

The definition of information measure is now generalised to deal with unequal probablities by *defining that the information associated with the symbol s, generated with probability p(s) by a zero memory source, is given by*

$$I(s) = \log \frac{1}{p(s)}$$

The *average* amount of information per symbol is then computed by the usual rule (equation 5.9)

$$\bar{I} = \sum_j I(s_j)p_j = \sum_j p_j \log \frac{1}{p_j}$$

where p_j is the probability of the jth symbol, s_j. This average information per symbol is called the source entropy, usually denoted H,

$$H = \sum_j p_j \log \frac{1}{p_j} \qquad \text{bits per symbol} \qquad 8.2$$

255

The statistical approach adopted here is seen to be inevitable when we reflect that a source generating a message must, as noted in Section 5.1, be classed as random. A deterministic source could never provide information, for its future output could be predicted with certainty from a knowledge of its past output. The entropy of a source is a measure of the average amount of uncertainty as to which symbol will be transmitted next. It measures the potential variety or 'surprise value' of the messages. A long message of N symbols carries NH units of information. Once again, however, it is stressed that the 'amount of information in a message' does not purport to measure the sense, truth, meaning, newsworthiness, or wisdom of the message or anything other than the statistical quantity NH which has been defined above for the special class of zero memory sources. In the words of C. E. Shannon, 'The semantic aspects of communication are irrelevant to the engineering problem.'

8.4 CODING SOURCE SYMBOLS

From the standpoint of information theory the communication process may be represented by the diagram of Fig. 8.2. The message is composed of symbols selected by the source from the source

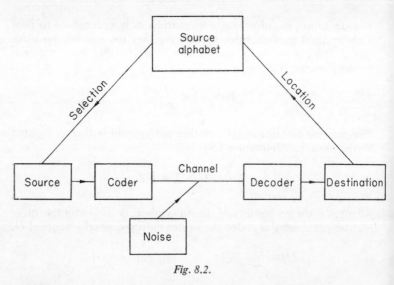

Fig. 8.2.

alphabet. The symbols are next coded into a form suitable for transmission (e.g. the ones and zeros of a teleprinter code or the dots and dashes of Morse code). Modulation may be included as part of the coding process. On receipt of the code symbols, the message is decoded, enabling the recipient to indentify the source symbols and reconstruct the message. Noise on the communication channel has the effect of making identification of the code symbols uncertain, so giving rise to errors in the decoded message. Knowledge of both the source alphabet and the code is assumed to exist at the destination, otherwise communication cannot take place.

Information sources were discussed in the preceding section and we turn now to examine the process of coding. The discussion will be restricted to binary coding, in which a fixed sequence of binary symbols is used to represent a source symbol. (The results may be extended to 'code alphabets' of more than two symbols.) Real-life situations in which the information source is truly zero-memory are hard to find, but we will suppose that the 'weather station' source of Section 8.3 fulfills this requirement. Possible binary codes for four alternative weather conditions are listed in Table 8.1, assuming first that all these are equally probable.

TABLE 8.1

S_j		P_j	Code 1	Code 2	Code 3
S_1	SUN	$\frac{1}{4}$	0	00	110
S_2	CLOUD	$\frac{1}{4}$	1	01	10
S_3	RAIN	$\frac{1}{4}$	01	10	0
S_4	SNOW	$\frac{1}{4}$	10	11	1110

$H = 2$ bits/symbol   REJECT $L = 2$ $L = 2{\cdot}5$

The most successful code is the one using the fewest binary code symbols (1's or 0's) to transmit a long message, since this represents a saving in time. In Table 8.1, code 1 appears to be the most economical from this point of view, until we notice that as there is no provision for 'space', it is impossible to decode uniquely a sequence such as 010, since this may mean 'sun, cloud, sun' or 'rain, sun,' or 'sun, snow'. This code is therefore rejected. In code 2 it has been tacitly agreed to use two code symbols per source symbol; there is then no need to code spaces and this code is quite acceptable. The objection is sometimes raised that zeros occurring at the beginning of a message would be missed, as we would not know that the message had started until a 1 occurred. This presupposes that a voltage level 0 is used to transmit that code symbol, but this is by no means necessary and any pair of distinguishable waveforms may used for the 1's and 0's. In any case, an agreed starting sequence such as 10101 would contribute negligibly to the length of a long message. Finally, code 3 is also seen to be uniquely decodable since this, in effect, uses 0 as a space symbol. In this code the individual lengths l_j, i.e. the number of code symbols per source symbol, are all different. The average code word length L is given by

$$L = \sum_j p_j l_j \qquad 8.3$$

and for code 3 this works out to be 2·5 code symbols per source symbol. In this example it is clear that code 2 is the more efficient. Notice that in this case $H = L = 2$ and no more efficient code can be devised.

Now, suppose the source symbol probabilities are different, as in the example presented by Table 8.2. The source entropy is now

TABLE 8.2

S_j		P_j	Code 2	Code 3
S_1	SUN	$\frac{1}{8}$	00	110
S_2	CLOUD	$\frac{1}{4}$	01	10
S_3	RAIN	$\frac{1}{2}$	10	0
S_4	SNOW	$\frac{1}{8}$	11	1110

$H = 1\cdot75$ bits/symbol		$L = 2$	$L = 1\cdot875$

somewhat smaller—in fact, it can be shown that for a given number of source symbols, the entropy is a maximum when they are all equally probable. This time, code 3 turns out to be the more efficient and the reason lies in the fact that the most frequently occurring source symbol, 'rain', has been given the shortest code word ($l_3 = 1$). while the less frequent symbols have been given longer code words,

258

This idea is not new, of course, and was used by Morse and Vail in constructing the Morse code which uses the shortest code element, the dot, to represent the most frequently occuring English letter, E, and much longer sequences of dots and dashes for the rarer letters. Notice that for both codes 2 and 3 we now have $L > H$.

There is clearly a need for a systematic method of devising economical codes. One method has been described independently by Shannon and by Fano. A different method has been given by Huffman, and his coding method will now be described. The source symbols are first ranked in order of their probabilities with the probabilies, p_j, listed alongside. This is illustrated in Fig. 8.3, for

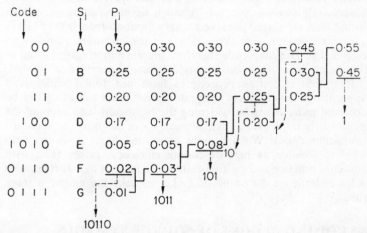

Fig. 8.3. *Huffman coding: tracing backwards from the second figure in the final column gives the code word for the symbol F, as shown. The other code words are obtained similarly (see text). For the source symbol probabilities chosen here, $H = 2\cdot31$ bits per symbol while the code gives $L = 2\cdot36$ binary symbols per symbol*

an alphabet of seven symbols. The bottom two figures are then added together and a second column written down, inserting the sum just obtained in the proper position. The process is continued until only two numbers remain. The upper number is labelled 0 and the lower one 1 and these numbers are now traced backwards, carrying the code words with them. Whenever a branch point is encountered, a 0 is added to the code word when the upper branch is followed and a 1 when the lower branch is followed. The code

259

words which finally arrive back at the first column are those used to code the individual symbols; in Fig. 8.3 they have been written in the extreme left-hand column. The rule '0 for up, 1 for down' could, of course, be reversed, giving a code with 1's and 0's interchanged. The reader is now invited to write down a random string of binary digits and see whether he can decipher this 'message' without ambiguity. There are no space symbols and the code words are not of equal length so that it is not immediately apparent that this code is uniquely decodable. The method of construction, however, ensures that *no code word has any other code word as a prefix* and it follows that, provided we have access to the beginning of the coded message, there is never any ambiguity in decoding. Notice that in this example $H = 2 \cdot 31$ bits/symbol while $L = 2 \cdot 36$ binary code symbols/symbol. A much simpler coding procedure would have employed the seven binary numbers from 000 to 110 in a constant-length code (like code 2 in the examples above) with $L = 3$. The Huffman code therefore presents an average saving in this example of $0 \cdot 64$ binary code symbols per source symbol.

If optimum coding procedures have not found widespread applications it is because adequate transmission rates can still be achieved more easily by widening the bandwidth and keeping the more easily implemented fixed-length codes (such as the standard teleprinter code). We are more concerned here, however, with what is *possible* as regards coding efficiency, rather than with common practice and we are beginning to suspect that H represents a lower limit on the value of L. Let us take the matter a stage further.

8.5 CODING BLOCKS OF SOURCE SYMBOLS. IDEAL CODES

Consider a very primitive source having just two symbols A and B, occurring with probabilities p and $1-p$, respectively. The source entropy is

$$H = p \log \frac{1}{p} + (1-p) \log \frac{1}{1-p} \quad \text{bits/symbol} \qquad 8.4$$

When $p = 0 \cdot 5$, $H = 1$ bit per symbol. For all other values of p the entropy is less than unity and tends to zero when $p \to 0$ or $p \to 1$. Of course when $p = 0$ the source can only generate a string of B's, with no chance of variety. When this is the case, and similarly for

$p = 1$, we are not surprised the information content turns out to be zero.

Now imagine a message sequence of N symbols generated by this source ($p \neq 0$ or 1). Counting up the number of A's and B's in the message, let us suppose that there are r A's and $N-r$ B's. The law of large numbers, Section 5.11, says that the probability that r lies *outside* the range

$$N(p-c) < r < N(p+c) \qquad\qquad 8.5$$

where c is a constant which may be chosen as small as we please, becomes negligible if N is sufficiently large. If $p = 0.7$, for example, then Fig. 5.1 shows that with $N = 1000$ we may confidently expect about 9995 out of every 10 000 messages to contain between 650 and 750 A's.

A scheme which codes the source symbol by symbol, as in the previous section, would provide a different code message for all the *possible* sequences of N symbols. Such a scheme is now seen to be wasteful, because when N is large, every message which the source produces will almost certainly contain about Np A's and $N(1-p)$ B's. For all practical purposes it will be sufficient to provide code sequences for these messages only. The number of possible sequences containing *exactly* Np A's is given by

$$W = \frac{N!}{(Np)!\,(N-Np)!}$$

The number of possible messages containing $Np \pm 1$, $Np \pm 2$, ... $Np \pm Nc$ of symbols A will also be given approximately by W when c is small. We may therefore estimate the number M of possible messages satisfying equation 3.5 as

$$M \approx 2NcW$$

This approximation is better expressed in logarithmic terms, since we may then simplify the expression for W using Stirling's approximation which states that, for large N and using natural logarithms

$$\log N! \approx N \log N - N$$

giving

$$\log W \approx N \log N - Np \log Np - N(1-p) \log N(1-p)$$
$$\approx -Np \log p - N(1-p) \log (1-p)$$

261

hence

$$\log M \approx \log 2c + \log N - Np \log p - N(1-p) \log (1-p)$$

The coding scheme now proposed is to list these M-messages and attach a binary code number to each. The minimum number of binary digits required for a message in this list is given by $\log_2 M$. The last equation above is independent of the base chosen for logarithms, hence the minimum number of binary code symbols used on average *per source symbol* is given by

$$L = \frac{\log_2 M}{N} \approx \frac{\log_2 2c}{N} + \frac{\log_2 N}{N} - p \log_2 p - (1-p) \log_2 (1-p)$$

This becomes exact as $N \to \infty$. The first two terms then vanish leaving

$$L = p \log_2 \frac{1}{p} + (1-p) \log_2 \frac{1}{1-p}$$

$$= H \text{ binary symbols per source symbol}$$

The 'code book' contains 2^{NL} messages and the method relies on the fact that, when N is large, the probability that the source will generate a sequence *not* on the list is negligible. This argument gives a new insight into the significance of the source entropy. With H measured in bits, a long sequence of N source symbols is almost certain to be one of a group containing $M = 2^{NH}$ different messages, the probability that the meassage is not one of these tending to zero for large N. This statement can be shown to be true for larger source alphabets than the primitive case considered above. The entropy is thus a measure of the potential variety in the messages which the source can produce.

The above coding method is quite impracticable. For example, if $N = 100$ and $H = 0.5$, we would need to list 2^{50} or about 10^{15} different messages in the code book. The following example, however, shows that even coding in pairs, rather than symbol by symbol, can yield a significant saving. Suppose that the results of a certain Saturday's football matches are to be transmitted, the list of games played being available at both source and destination. The information source runs down the list, announcing H (home win), A (away win) or D (match drawn) for each game in turn. This source is certainly zero-memory, and a typical message might begin

HAHADHHAAHHAAHDHAHHA ...

We shall assume that from previous experience the probabilities of the source alphabet symbols are known to be $p(\text{H}) = 0.5$, $p(\text{A}) = 0.4$, and $p(\text{D}) = 0.1$. The source entropy is then $\text{H} = 1.36$ bits/symbol. Now imagine that the source provides 5 symbols per second so that information is being generated at the average rate 6.8 bits/second, while the only communication link available can send up to 7 ON or OFF pulses per second. Can we code the source in such a way that the channel can cope with symbols arriving at 5 per second? First we try encoding symbol by symbol (Table 8.3).

TABLE 8.3

S_j	P_j	Code 1	Code 2
H	0·5	00	0
A	0·4	01	10
D	0·1	10	11

The first code requires two code symbols for each source symbol, $L = 2$, and thus 10 binary symbols per second would need to be transmitted using this code. The second code (a Huffman code) uses an average length $L = 1.5$ so that an average of 7.5 binary symbols per second would be produced. Although more efficient than the primitive code 1, this still exceeds the channel capability. The next stage is to code in pairs. For a zero memory source the probability of occurrence of any pair of symbols may be computed from equation 5.7. These are listed in Table 8.4, together with a Huffman code, whose construction is left as an exercise for the reader.

TABLE 8.4

S_j	P_j	Code 3	l_j
HH	0·25	01	2
HA	0·20	10	2
AH	0·20	11	2
AA	0·16	001	3
HD	0·05	00000	5
DH	0·05	00001	5
AD	0·04	00011	5
DA	0·04	000100	6
DD	0·01	000101	6

$$L = \sum_j l_j p_j = 2.78$$

This code uses an average of 2·78 binary symbols for each pair of source symbols, or 1·39 binary symbols per source symbol. (cf. $H = 1·36$ bits per symbol). The source symbols arrive at 5 per second so that this code will produce an average of 6·95 binary symbols per second. This is just within the capability of the channel. (A temporary store would be needed to smooth out fluctuations caused, for example, by a string of D's but the *average* rate is satisfactory.) A further small improvement would be possible by encoding longer source words but, as we saw above, *it is impossible to devise a code for which $L < H$.*

This important result is embodied in a fundamental theorem due to Shannon, which for the special case of a binary channel may be expressed as follows:

Theorem: Let a source have entropy H bits per symbol and a binary channel be capable of transmitting C binary digits per second. Then it is possible to encode the output of the source in such a way as to transmit at the average rate of $(C/H - \delta)$ symbols per second over the channel, where δ is arbitrarily small. It is not possible to transmit at an average rate greater than C/H.

It is now possible to define the efficiency, η, of a code used in conjunction with a source of entropy H, as

$$\eta = \frac{H}{L} \qquad 8.6$$

Except in special cases (e.g., code 2, Table 8.1), no code can have unit efficiency. Some of the code symbols used in coding a message will be strictly unnecessary. The fraction which is unnecessary is termed the *redundancy* of the code, measured as $1 - \eta$. Notice that an ideal code is suited to only one set of source symbol probabilites. In this respect the coding operation may be described as one of 'matching' the information source to the communication channel in such a way as to achieve the maximum possible rate of information transfer. This maximum rate is the *information capacity* of the channel, about which more will be said in the next chapter.

8.6 SIMPLE ERROR-CORRECTING CODE

If the 'football results' message sequence suggested on page 262 is encoded using code 3 above and then just the first code symbol is changed from 1 to 0, the result is catastrophic, for decoding then produces a quite different sequence of source symbols. This behaviour is typical of low-redundancy codes and makes them unsuitable for

transmission over a channel in which the amount of noise is sufficient to produce a significant probability of error in the received
code symbols. Hamming and others have evolved special codes in
which errors may be detected and corrected. One of the simplest examples is a 'parity check' code which works as follows.

The coded message is divided into 'words' containing, for example, 16 binary symbols. One such code word might be

$$0 1 1 0 1 0 1 0 0 1 1 1 0 0 0 1$$

This sequence is written as a 4×4 array and 'check' digits are added
to give one extra row and one extra column in such a way that
the total number of 1's in each row and column is *even*. The result
is the 5×5 array

$$
\begin{array}{c|c}
0110 & 0 \\
1010 & 0 \\
0111 & 1 \\
0001 & 1 \\
\hline
1010 & 0 \\
\end{array}
$$

The sequence transmitted is then

$$0 1 1 0 0 1 \overset{\circ}{0} 1 0 0 0 1 1 1 1 0 0 0 1 1 1 0 1 0 0$$

Let us assume that the starred symbol is changed by the noise from
0 to 1. The 5×5 array set up at the receiver is now

$$
\begin{array}{l}
01100 \\
11100 \leftarrow \\
01111 \\
00011 \\
10100 \\
\uparrow
\end{array}
$$

A parity check reveals that an error has occurred in the second row
and in the second column. The symbol in error is thus located,
corrected, and the original 16-symbol code word is recovered intact.

Two errors in the received 5×5 array could be detected but not
located and corrected. If the error probability is w per symbol then
the chances of a single error and of two errors are given by

$$p(1) = 25w(1-w)^{24}$$

$$p(2) = \frac{25!}{2! \, 23!} w^2(1-w)^{23}$$

Signals and Information

Thus if $w = 10^{-4}$ we have, approximately, $p(1) = 2 \cdot 5.10^{-3}$ and $p(2) = 3.10^{-6}$.

In transmitting a long message containing 10 000 words of 25 binary symbols, an average of about 25 errors would be satisfactorily corrected in the way described above, while the chances are about 33 : 1 against the complete message not being recoverable on account of more than 1 error occurring in one of its 25-symbol words. Without the check symbols, 160 000 binary symbols would have been transmitted and about 16 uncorrectable errors would almost certainly have occurred. Randomly occurring errors with low error rates are threfore successfully defeated using this scheme. Larger error rates, or errors occurring in bursts, require more complex schemes which are beyond the scope of the present discussion.

The important feature to notice is that an error-correcting code is necessarily *redundant*. The way to combat noise is to use a (judiciously) redundant code. It has been estimated that English text is about 50% redundant. It is this which makes it possible to hold a conversation over a noisy telephone line or to detect and correct the typographical errors which sometimes occur in books and newsprint.

Further Reading

SHANNON, C. E., 'A mathematical theory of communication', *Bell Syst. tech. J.* **27**, 379–423 and 623–658 (1948)

PIERCE, J. R., *Symbols, Signals and Noise* (Hutchinson, London, 1962)

WOODWARD, P. M., *Probability and Information Theory, with Applications to Radar* (Pergamon, London, 1953)

ABRAMSON, N. M., *Information Theory and Coding* (McGraw-Hill, New York, 1963)

SCHWARTZ, L. S., *Principles of Coding, Filtering, and Information Theory* (Cleaver-Hume, London, 1963)

HUFFMAN, A., 'A method for the construction of minimum redundancy codes', *Proc. Inst. Radio Engrs* **40**, 1098, 1952.

PROBLEMS

1. Good picture quality is still maintained when the brightness levels in a television picture are quantised to 64 levels. If this resolution is possible at 500 points along each line of a 625-line picture, what is the potential information content of one picture?

2. Sketch the function $H = p \log \dfrac{1}{p} + (1-p) \log \dfrac{1}{1-p}$.

3. Five source symbols have the probabilities $p_1 = 0 \cdot 40$, $p_2 = 0 \cdot 25$,

$p_3 = 0.20$, $p_4 = 0.10$, $p_5 = 0.05$. Show that the source entropy is 2·04 bits/symbol (0·614 hartleys/symbol). Using Huffman's method, devise a binary code for this source and evaluate the code efficiency.

4. A zero-memory source has entropy H bits per symbol. Prove that the entropy per 'word' of two symbols is given by

$$H_2 = \sum_{jk} p_j p_k \log \frac{1}{p_j p_k} = 2H$$

Verify that the 'football results' source of Section 8.5 obeys this equation.

5. *Shannon coding.* In this method the source symbols are again listed in order of probability with the most probable symbol, s_1, at the head of the list. Cumulative probabilites P_j are then defined, such that P_j is the sum of the individual probabilities Σp_k for $k < j$, i.e., the sum of those lying *above* the symbol s_j (set $P_1 = 0$). The code for s_j formed by expanding P_j as a binary fraction, continuing to l_j places, where

$$\log_2 \frac{1}{p_j} \leq l_j < 1 + \log_2 \frac{1}{p_j}$$

i.e.
$$\frac{p_j}{2} < 2^{-l_j} \leq p_j$$

Show that the code word for s_j will differ from that for all succeeding symbols in one or more of its l_j places and hence that unique decoding is possible.

Show also from the first inequality above that the average code word length, L, satisfies the inequality

$$H \leq L < H + 1$$

and verify this for the source of question 3.

6. From the result of Problem 4 and the last inequality in Problem 5 show that when coding pairs of source symbols, using Shannon's method, the average number of binary symbols per source symbol satisfies the inequality

$$H \leq L < H + \tfrac{1}{2}$$

and hence that, by coding sufficiently long sequences, $L \to H$.

7. The noise on a certain channel is such that the probability of a binary symbol being received in error is $w = 0.1$. In an effort to overcome the noise, each binary symbol is transmitted five times. Show that the probability of error is now $8.56.10^{-3}$.

8. The binary symbols issuing from a teleprinter or a computer output device are sometimes called bits. Is this strictly correct?

Chapter 9

Communication Channels

9.1 INTRODUCTION

It was seen in the preceding section that by using a redundant code it is possible to reduce the probability of error in the received message. A somewhat crude method would be to repeat each binary symbol three times, and then so long as the noise does not alter more than one binary symbol in each group of three, the error can be detected, leaving no doubt at the receiver as to which symbol was intended. A heavy price has been paid, however, since the speed of transmission has been cut to one-third. Even then we cannot be sure the noise has not defeated us by changing more than one symbol in some of the groups of three. To make the chance of a message error approach zero it would appear necessary to increase the redundancy indefinitely, the speed of transmission (i.e., the rate of information transfer) tending to zero.

Shannon, however, has shown that the conclusion we have just reached is by no means true. He proved that it is possible to send messages over a noisy channel *with as small a probability of error as desired* while still transmitting at a finite speed, provided no attempt is made to send information faster than a certain rate known as the *channel capacity* (measured either in bits per second or bits per code symbol). In this chapter we shall examine this very remarkable statement and look briefly at some of its consequences.

9.2 CHANNEL CAPACITY

Returning for a moment to the noise-free situation, consider the binary signalling arrangement of Fig. 9.1(a). This channel is capable of transmitting one binary symbol in each time interval τ. *If this noiseless channel is used in conjunction with an ideal code* then each

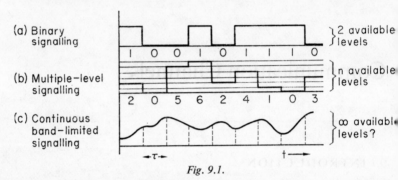

(a) Binary signalling — 2 available levels

1 0 0 1 0 1 1 1 0

(b) Multiple-level signalling — n available levels

2 0 5 6 2 4 1 0 3

(c) Continuous band-limited signalling — ∞ available levels?

Fig. 9.1.

binary symbol carries one bit of information, so that information may be carried at the rate

$$C = 1 \quad \text{bit per symbol}$$

or

$$C = 1/\tau \text{ bits per second}$$

This is the *channel capacity*, and it is important to observe that the capacity is only utilised when using an ideal code.

The code alphabet may include more than two symbols and, although only binary codes have been cosidered here, it is possible to devise codes which employ larger code alphabets. The code symbols are transmitted using a different waveform for each. One simple method is to use a different voltage level for each code symbol as illustrated in Fig. 9.1(b). Another method might be to transmit $\cos[\omega(t - m\tau) + \phi_k]$ in the interval $m\tau < t < (m+1)\tau$ and use different phase angles ϕ_k for the different code symbols. (This technique, known as phase shift keying, has the advantage of maintaining constant transmitter power. It can, of course, be used for binary signalling.) If n code symbols are available, then maximum information transfer occurs when they are all equally probable and the channel capacity becomes

$$C = \log_2 n \quad \text{bits per symbol}$$

or

$$C = \frac{\log_2 n}{\tau} \quad \text{bits per second}$$

The extent to which channel capacity is reduced by noise will be discussed below. At this stage, however, we may observe that the continuous waveform of Fig. 9.1(c), band-limited to F Hz, may be sampled $2F$ times per second. The capacity then takes the form

$$C = 2F \log_2 n \text{ bits per second} \qquad 9.1$$

where n is the number of 'available levels'. For the discrete channels considered above, n is finite, but for this *continuous* channel n is infinite and, therefore, so is the channel capacity. When noise is present, however, only a finite number of levels will be distinguishable with any certainty, so we may still expect the capacity of a continuous channel to have the form of equation 9.1. The proper choice for n will be the concern of Section 9.6.

9.3 DISCRETE CHANNELS WITH NOISE

When noise is present on the waveforms (a) or (b) of Fig. 9.1, the receiver can do no better than take an average $\bar{v}$ of the signal received in a given interval τ and announce that the most likely symbol transmitted was that corresponding to the level v_k nearest to $\bar{v}$. We will denote the selection of the jth symbol for transmission by x_j, while the receiver's response, 'kth symbol', will be denoted y_k. In the noiseless case, $k = j$ with unit probability. For a noisy channel however, the transmission of x_j may result in the reception of any y_k. The probability that y_k will be received when x_j is transmitted is written $p(y_k \mid x_j)$, following the notation of Section 5.3. The probabilites $p(x_j)$ which govern the selection of symbols for transmission are assumed to be known, while the $p(y_k \mid x_j)$ may be determined by observation over a long period or else computed from the statistics of the noise. The reader may find it helpful to imagine the situation as depicted in Fig. 9.2. Here one of a row of buttons, x_j, is depressed at the transmitter. This may result, with known probabilities, in any of the lamps y_k flashing at the receiver.

From the $p(x_j)$ and $p(y_k \mid x_j)$ we may compute the probabilities of the joint events (x_j, y_k)

$$p(x_j, y_k) = p(x_j) p(y_k \mid x_j) \qquad 9.2$$

271

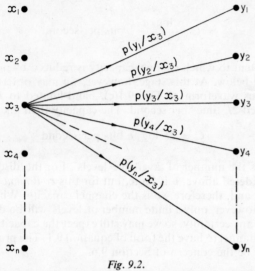

Fig. 9.2.

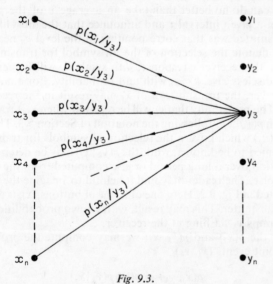

Fig. 9.3.

(see Section 5.3). There are also the probabilities with which symbols are received

$$p(y_k) = \sum_j p(x_j, y_k) \qquad 9.3$$

and finally the 'backward probabilities' $p(x_j|y_k)$ which give the probability at the receiver that x_j was sent, given that y_k is received (Fig. 9.3)

$$p(x_j|y_k) = \frac{p(y_j, y_k)}{p(y_k)} \qquad 9.4$$

The entropy of the output from the encoder (assuming this to behave as a zero memory source) may be written

$$H(X) = \sum_j p(x_j) \log \frac{1}{p(x_j)} \qquad 9.5$$

which measures the average uncertainty (per symbol) as to which symbol will be transmitted next. When y_k appears at the receiver, this entropy changes to

$$H(X|y_k) = \sum_j p(x_j|y_k) \log \frac{1}{p(x_j|y_k)}$$

Averaging over all the y_k's we may define a quantity

$$H(X|Y) = \sum_{j,k} p(y_k) p(x_j|y_k) \log \frac{1}{p(x_j|y_k)}$$

$$= \sum_{j,k} p(x_j, y_k) \log \frac{1}{p(x_j|y_k)} \qquad 9.6$$

which measures the average remaining uncertainty per symbol as to which was the transmitted symbol, after a symbol has been received. In the noiseless situation $p(x_j|y_k) = 1$ for $j = k$ and is zero otherwise. It is easily seen that $H(X|Y) = 0$ in the noiseless case. When noise is present, the average uncertainty which remains after receipt of a symbol, measured by $H(X|Y)$, is called the *equivocation*. The initial uncertainty $H(X)$ minus the final uncertainty

273

$H(X|Y)$ is the average amount of information per symbol transmitted over the channel. This is commonly denoted $I(X; Y)$

$$I(X; Y) = H(X) - H(X|Y) \qquad 9.7$$

$$= \sum_j p(x_j) \log \frac{1}{p(x_j)} - \sum_{j,k} p(x_j, y_k) \log \frac{1}{p(x_j|y_k)}$$

$$= \sum_{j,k} p(x_j, y_k) \left[\log \frac{1}{p(x_j)} - \log \frac{1}{p(x_j|y_k)} \right]$$

$$= \sum_{j,k} p(x_j, y_k) \log \frac{p(x_j|y_k)}{p(x_j)}$$

$$= \sum_{j,k} p(x_j, y_k) \log \frac{p(x_j, y_k)}{p(x_j)\, p(y_k)} \qquad 9.8$$

where in the last line we have used equation 9.4. The symmetry of equation 9.8 with respect to x_j and y_k has led to the term 'mutual information' for $I(X; Y)$.

For a noise-free channel we may write

$$p(x_j, y_k) = p(x_j) = p(y_k) \qquad k = j$$
$$p(x_j, y_k) = 0 \qquad k \neq j$$

in which case the right-hand side of equation 9.8 reduces to the entropy $H(X)$ as required. The equivocation is then zero. On the other hand, when the noise power is very large, the probability of 'receiving' y_k becomes independent of the transmitted symbol x_j. The lamps at the receiver flash at random, in response to the noise alone, and no information is transferred. The events x_j and y_k are then statistically independent so that $p(x_j, y_k) = p(x_j)\, p(y_k)$ and, by equation 9.8, $I(X; Y) = 0$.

Thus, from a consideration of the entropies, the above analysis shows that information is sent over a noisy channel at a definite rate per symbol, $I(X; Y)$ in spite of the noise. How this information can appear in the form of error-free messages will be discussed later, but first we may notice that the mutual information depends not only on the noise—which governs the probability of mistaking one code symbol for another, the $p(y_k|x_j)$—but also upon the transmitted symbol probabilities, $p(x_j)$. For a given channel $I(X; Y)$ will be a maximum for one particular set of values of the $p(x_j)$. This maximum achievable rate of information transfer is the *capacity* of the noisy channel, thus

$$C = I(X; Y)_{\max} \quad \text{bits per symbol} \qquad 9.9$$

The problem of maximising the mutual information with respect to the $p(x_j)$ is a very difficult one in the general case. Obtaining the optimum values for the $p(x_j)$ provides a further problem which is essentially one of coding. It is seen that, as in the noiseless case, the channel capacity can only be achieved by using an optimum code. The easiest case to consider, which fortunately is also the most important one, is that of the noisy binary channel.

9.4 NOISY BINARY CHANNEL

Suppose two symbols, which will be labelled A and B rather than 1 and 0 for clarity of notation, are transmitted with probabilities $\frac{3}{4}$ and $\frac{1}{4}$, respectively. Let us also assume that the chance of an error in transmission is $\frac{1}{4}$, as shown in Fig. 9.4. Thus we have,

$$p(x_A) = \tfrac{3}{4} \qquad p(y_A \mid x_A) = p(y_B \mid x_A) = \tfrac{3}{4}$$
$$p(x_B) = \tfrac{1}{4} \qquad p(y_B \mid x_A) = p(y_A \mid x_B) = \tfrac{1}{4}$$

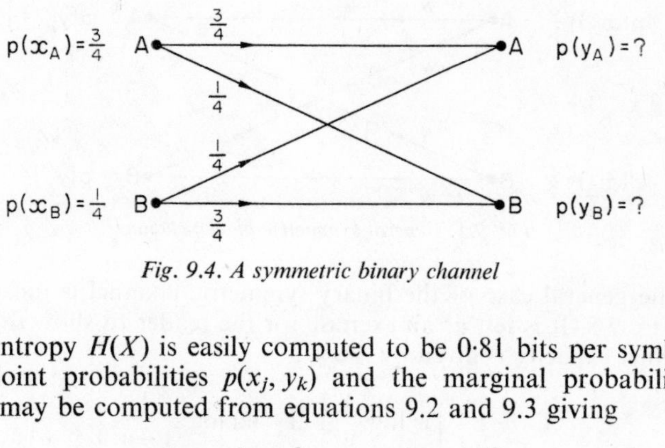

Fig. 9.4. A symmetric binary channel

The entropy $H(X)$ is easily computed to be 0·81 bits per symbol. The joint probabilities $p(x_j, y_k)$ and the marginal probabilities $p(y_k)$ may be computed from equations 9.2 and 9.3 giving

$$p(x_A, y_A) = \tfrac{9}{16} \qquad p(y_A) = \tfrac{5}{8}$$
$$p(x_A, y_B) = \tfrac{3}{16} \qquad p(y_B) = \tfrac{3}{8}$$
$$p(x_B, y_A) = \tfrac{1}{16}$$
$$p(x_B, y_B) = \tfrac{3}{16}$$

Application of equation 9.8 now gives

$$I(X; Y) = 0.14 \text{ bits per symbol}$$

and the equivocation works out at $H(X|Y) = 0.67$ bits per symbol. The equivocation, representing the information destroyed by the noise, is quite large in this example but about 0.14 bits per symbol are still sent over the channel.

In this example the probabilities that A will be mistaken for B and vice versa have been deliberately chosen to be the same. For this *symmetric binary channel* it comes as no great surprise to learn that the mutual information is a maximum when $p(x_A) = p(x_B) = 0.5$. With these probabilities for the transmitted symbols and with the error rate as in the example above, we have

$$p(x_A, y_A) = p(x_B, y_B) = \tfrac{3}{8} \quad p(y_A) = p(y_B) = \tfrac{1}{2}$$
$$p(x_A, y_B) = p(x_B, y_A) = \tfrac{1}{8}$$

giving $I(X; Y) = 0.19$ bits per symbol. This is the largest mutual information attainable for the error rates quoted and thus the channel capacity is 0.19 bits per symbol.

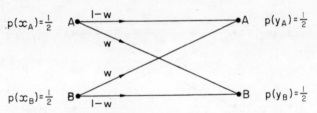

Fig. 9.5. General symmetric binary channel

The general case of the binary symmetric channel is illustrated in Fig. 9.5. It is left as an exercise for the reader to show that the channel capacity is given by

$$C = 1 - \left[w \log \frac{1}{w} + (1-w) \log \frac{1}{1-w} \right] \qquad 9.10$$

The expression within the square brackets is the equivocation, and the similarity with the entropy expression, equation 8.4, is worth noting. One way of interpreting equation 9.10 which is sometimes found helpful is as follows. Suppose an external observer can see both the transmitted symbols and the received symbols. He will therefore know when the noise has produced an error and we will imagine he is equipped with a noiseless channel to the receiver, so that he can tell the receiver about the errors by flagging 1 when an error has been made and 0 otherwise. The receiver is now getting information via two channels (i) by way of the noisy channel connecting him to the transmitter and (ii) by way of the noiseless correcting channel which links him to the external observer. The combination is clearly equivalent to noiseless communication with the source, for no net errors are now being made. Information is therefore received at one bit per symbol. Some of this, however, is coming through the correcting channel. The observer is flagging 1 with probability w and 0 with probability $1-w$. He is in fact an information source providing

$$H_{\text{Ext}} = w \log \frac{1}{w} + (1-w) \log \frac{1}{1-w}$$

bits of information per transmitted symbol. The remaining information, $1-H_{\text{Ext}}$, must be coming over the noisy channel. This agrees with equation 9.10 and gives the information arriving via the noisy channel whether the observer is there or not.

9.5 ERROR-FREE BINARY MESSAGES

Rigorous proofs are available to show that error-free messages can be sent over a noisy channel in such a way as to transfer information at a rate which can be made arbitrarily close to the channel capacity. These proofs are somewhat long and involved, and, as an alternative, the following pictorial argument is offered as a sufficiently convincing demonstration as regards the binary symmetric channel. We shall assume a small error probability, w, though the final result is quite general.

Fig. 9.6 shows a three-dimensional Cartesian reference frame in which the eight points 000 to 111 form the corners of a unit cube. These eight points may be used to represent the eight possible binary words which are composed of just three binary symbols. Suppose one of these words is transmitted over a noisy channel and one symbol is erroneously received. For example, 110 might be sent and either 010, 100 or 111 received. Any of these altered words may be reached by taking one step along an edge of the cube. The sequences containing two errors may be reached by taking two

277

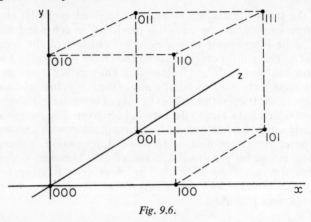

Fig. 9.6.

steps. Extending this picture to longer sequences of N binary symbols, we may imagine a unit 'cube' in an N-dimensional hyperspace, each corner of which represents one of the 2^N possible sequences of N symbols. A sequence which differs from a given sequence in just one symbol may be reached by taking one step along an edge, those differing in two places lie within two steps, and so on. When a given sequence is transmitted over a noisy channel, the effect of the noise is to change the sequence to a different one lying r steps away, where r is the number of errors in N symbols. The number of steps is called the *distance* between the sequences.

The number of sequences lying at a distance r from a given sequence is just the number of ways, W, of choosing r from N

$$W = \frac{N!}{r!\,(N-r)!}$$

and it is an intriguing feature of the hypercube that when N is large and $r \ll N$ (an error rate of less than 10%, say), then almost all the sequences lying *within* a distance r actually lie exactly r steps away. This is not difficult to prove, but an example will suffice. Suppose $N = 1000$ and $r = 3$ then using the expression for W above, about 166 million sequences lie exactly 3 steps away from some given sequence, while only about 0·5 million lie 2 steps away and a mere 1000 lie one step away. For not too large error rates, therefore, the expression above for W may be used to compute the number of different sequences which lie *within* a given distance r.

278

Invoking the law of large numbers (Section 5.11), the probability that a received sequence of N symbols will contain more than $N(w+c)$ errors can be made as small as desired by making N sufficiently large, no matter how small a value is (initially) chosen for c. Thus, by using long sequences to code the messages, we may be virtually certain that the received sequence lies within a distance r of that transmitted where $r = N(w+c)$ and c is arbitrary. With the aid of Stirling's approximation we then have

$$\log W = \log N! - \log N(w+c)! - \log [N - N(w+c)]!$$
$$\approx -N(w+c) \log (w+c) - N(1-w-c) \log N(1-w-c)$$

but we may choose $c \ll w$, hence

$$\log W \approx -Nw \log w - N(1-w) \log (1-w)$$
$$\approx N \left[w \log \frac{1}{w} + (1-w) \log \frac{1}{1-w} \right]$$

In principle at least, a number of messages may now be coded for error-free transmission in the following way. Some particular sequence of binary digits is selected (arbitrarily) and labelled 'Message 1'. At the receiver, this sequence *and the W other sequences lying around it* are also labelled 'Message 1'. For large N it is now virtually certain that when Message 1 is transmitted, the sequence received will lie on this particular 'patch' of the hypercube. The chance that it lies outside may be made as small as we please by making N large enough. We may do the same for another message, 'Message 2' and provided 'patch 1' and 'patch 2' do not overlap, then no received sequence can leave us guessing as to which message was sent. The largest number, of messages which can be handled in this way will be given by the number of patches when the patchwork completely envelopes the hypercube. There are 2^N corners to be covered so that an upper limit for M is given by

$$M = \frac{2^N}{W}$$

or

$$\log_2 M = N - \log_2 W = N \left[1 - w \log \frac{1}{w} - (1-w) \log \frac{1}{1-w} \right]$$

An information source which generates the M messages will have maximum entropy when these are all equally probable. In this case

19

the information conveyed per message is simply $\log_2 M$ bits. This information is now being carried by N binary symbols, so the (maximum) information per binary symbol is

$$C = \frac{\log_2 M}{N}$$

$$= 1 - w \log_2 \frac{1}{w} - (1-w) \log_2 \frac{1}{1-w} \quad \text{bits per symbol}$$

in agreement with equation 9.10.

A non-redundant code designed for M equally likely messages would employ $\log_2 M$ binary symbols per message. The code above uses N binary symbols per message and thus the code efficiency is

$$\eta = \frac{\log_2 M}{N}$$

and the *redundancy* is

$$1 - \eta = w \log_2 \frac{1}{w} + (1-w) \log_2 \frac{1}{1-w}$$

The redundancy is therefore precisely equal to the equivocation of the channel. At the end of the previous chapter the comment was made that it is redundancy in a code which makes communication possible in the presence of noise. The triumph of Shannon's theory lies in his demonstration of the existence of codes with just sufficient redundancy to permit error-free message transmission. Unfortunately, the huge number of possible code sequences when N is large makes the coding method suggested above quite impracticable. Shannon's theorem itself is essentially an existence theorem and does not provide a technique for devising ideal, minimum-redundancy codes for reliable communication. It gives a clue, however, that the best way to tackle the problem is to encode using long blocks and to introduce a controlled amount of redundancy, sufficient to ensure unambiguous identification of the code words. This is much better than simply repeating each symbol many times and is in fact the technique of the elementary parity-check code of Section 8.6. Systematic error correcting codes have been devised by Hamming and others but the reader is referred to more specialised texts for a description of these.

9.6 CONTINUOUS CHANNELS WITH NOISE

In the previous section it was found helpful to represent a binary code word, and therefore in effect a signal like that of Fig. 9.1(a), by means of a single point in a many-dimensional space. In Chapter 2 (Section 2.7.) a similar representation was suggested for continuous waveforms, an idea which is explained more fully in Appendix 3. We know that a section of waveform band-limited to F Hz and lasting for T seconds may be specified in terms of its $2FT$ samples. The sampling functions, of the form sinc $2\pi Ft$, are mutually orthogonal. It follows that the sample values v_k of such a signal may be used as the coordinates of a point in a $2FT$-dimensional space and that this point completely defines the signal.

By analogy with two- and three-dimensional spaces, the distance R of the point from the origin is given by

$$R^2 = \sum_k v_k^2$$

In Section 2.16 it was shown that the mean power P of the signal is given by

$$P = \frac{1}{2FT} \sum v_k^2$$

Hence

$$R = \sqrt{(2FTP)}$$

If the waveform is a random waveform, being either noise or signal or signal plus noise, then R is a random variable. However, when $2FT$ is very large (law of large numbers again) the value of R is almost certain to be almost exactly $\sqrt{(2FTP)}$ where P is now the variance of the sample voltages. It follows that a message waveform lasting for time T, taken from an ensemble of signals of mean power S, is almost certain to correspond to a point which lies very close to the surface of a hypersphere whose radius is $\sqrt{(2FTS)}$. Let such a signal be received together with added noise of mean power N. The received power is $S+N$ and so the received waveform must be represented by a point lying close to the surface of a second hypersphere whose radius is $\sqrt{[2FT(N+S)]}$. In Fig 9.7 the point Q represents a received waveform. The signal waveform which would have been received in the absence of noise lies somewhere at a distance of almost precisely $\sqrt{(2FTN)}$ from Q and close to the surface of the inner hypersphere. That is to say, the message

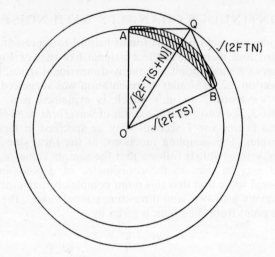

Fig. 9.7.

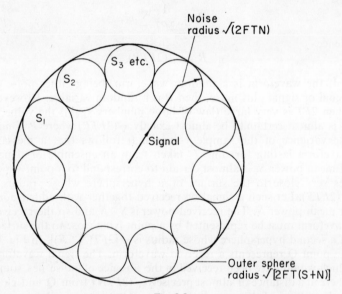

Fig. 9.8.

signal waveform defines a point which must lie somewhere in the shaded region AB.

Messages may be coded for error-free transmission by using a selection of code waveforms represented by points distributed around the inner hypersphere. When waveform Q is received there will be no ambiguity about the transmitted waveform, provided only one code waveform lies within the 'uncertainty region' AB. Fig. 9.8 shows message waveform points S_1, S_2, S_3, etc. It will clearly be possible in principle to distinguish between two transmitted signals, provided that the little hyperspheres of radius $\sqrt{(2FTN)}$, constructed about their respective signal points, do not overlap. The number of distinguishable signals which may be transmitted certainly cannot exceed the number of 'noise spheres' which can be packed around the inner surface of the large hypersphere of radius $\sqrt{[2FT(S+N)]}$.

The area of a circle (a two-dimensional 'sphere') of radius r is proportional to r^2, the volume of a three-dimensional sphere is proportional to r^3 and it can be shown that the volume of a hypersphere in n dimensions is proportional to r^n. When the dimensionality is large, the volume of a hypersphere lies almost entirely in its outermost region. (This property is similar to one noticed for the hypercubes in the previous section.) If kr^n is the volume within radius r then the volume between radii ϱ and r is given by

$$V = k[r^n - \varrho^n]$$
$$= kr^n[1 - (\varrho/r)^n]$$

and if n is sufficiently large (we are assuming $2FT \gg 1$) the second term in the square brackets becomes negligible, however close ϱ is to r. Thus the volume available to the little hyperspheres is essentially the whole volume of the large hypersphere, namely

$$V = k\{\sqrt{[2FT(S+N)]}\}^{2FT}$$

The number of 'noise spheres' which may be packed inside cannot exceed the ratio of the volumes, namely

$$M = \frac{\{\sqrt{[2FT(S+N)]}\}^{2FT}}{[\sqrt{(2FTN)}]^{2FT}}$$
$$= \left(1 + \frac{S}{N}\right)^{FT}$$

This is the maximum number of distinguishable message waveforms, and so the information per message is $\log_2 M$ bits. Dividing $\log_2 M$ by the time T spent in transmission gives the channel capacity in bits per second,

$$C = F \log_2 \left(1 + \frac{S}{N}\right) \quad \text{bits per second} \qquad 9.11$$

This, though not the rather primitive derivation, is Shannon's famous result giving the capacity of a continuous channel of bandwidth F perturbed by white Gaussian noise. The argument above arrives at this expression as an upper limit. Proper analysis shows that the channel capacity in fact tends to this limit when the product FT is sufficiently large. Different messages, or message sequences, are to be coded onto the M waveforms, which are now long complicated signals looking much like noise waveforms. These code signals are assumed to be known at the receiver, where a received waveform is compared with each in turn. The one selected is that giving the best least-squares fit (i.e., the nearest code signal in the hyperspace) and, with M as given above, the chance of making the wrong choice can be made vanishingly small.

The huge complexity of the equipment which would be required to implement such a scheme puts it far beyond the reach of practical possibility, just as the coding scheme suggested in Section 9.5 was too involved to be usable except in a 'thought experiment'. Practical communications systems are perforce inefficient when compared with the ideal of equation 9.11. The result stands, nevertheless, as one of great theoretical significance.

9.7 MORE ABOUT CONTINUOUS CHANNELS. JOULES PER BIT

To get some idea of the numbers involved, let us use equation 9.11 to calculate the maximum possible rate of information transfer for some familiar channels. A 100-words-per-minute telegraph channel is usually allotted a bandwidth of 170 Hz or so on an underground cable and, it we assume that satisfactory operation requires a signal-to-noise ratio of 20 dB, then equation 9.11 gives a capacity $C = 1000$ bits per second, approximately. A conventional telegraph transmission falls far short of this capability. A telephone channel with a 3 kHz bandwidth and $S/N = 30$ dB, corresponding roughly to

commercial speech standards, has a capacity of some 30 000 bits
per second. At the other end of the scale, a television link, with a
5 MHz bandwidth and $S/N = 45$ dB would have a capacity of
about $7.5 \cdot 10^7$ bits per second. It has been estimated that the entropy
of English text is about 1 bit per letter, or 5 bits per word. If we
reckon that a reader who is concentrating hard can read English
prose at around 300 words per minute, then we can get an estimate of the rate at which a human being can receive and process
information. This works out at around 25 bits per second. It would
be dangerous to assume that human performance can really be
measured in this simple way, but it is none the less sobering to
speculate that a television channel, if used as a communication
link at maximum efficiency, would be capable of sending enough
information in one second to keep most of us occupied for a month.

Equation 9.11 combines in a single expression two basic features
of electrical communication systems, both of which have caught
our attention in earlier chapters. The first, which was noted in Chapter 2, is that the speed of transmission is directly proportional to the
available bandwidth. The second feature is the ability to exchange
bandwidth for signal-to-noise ratio, while maintaining the same
channel capacity. Shannon's law shows that if the noise power increases, so that S/N becomes smaller, then the channel capacity can be
restored by increasing the bandwidth. In Chapter 4 it was seen that
this is a characteristic of wideband modulation systems such as
f.m., p.p.m., and p.c.m. These modulation techniques by no means
form ideal coding systems so that they do not afford all the improvement suggested by equation 9.11; it is just that they are sufficiently
flexible to take advantage of what is now seen to be a fundamental
property of information-bearing signals in noise.

The received energy expended on each bit of information may
also be computed from equation 9.11 as follows. A signal lasting
for time T uses ST joules to send CT bits of information. The
cost per bit is therefore

$$E = \frac{S}{C} = \frac{S}{F \log_2 (1 + S/N)}$$

$$= \frac{S \log_e 2}{F \log_e (1 + S/N)}$$

$$= \frac{0.693}{F} \frac{S}{\log_e (1 + S/N)} \quad \text{joules per bit}$$

285

where logarithms are now to base e. Somewhat surprisingly, perhaps, this function is seen to take its minimum value when $S/N \to 0$. When $S/N \ll 1$ the above expression reduces to

$$E_{\min} = \frac{0.693N}{F} \quad \text{joules per bit}$$

A receiver of bandwidth F, matched to a source at temperature T, will receive, as we saw in Chapter 6, a mean noise power kTF watts. Thus the minimum energy which a signal must deliver in providing 1 bit of information, against a background of noise at temperature T, is given by

$$E_{\min} = 0.693kT \quad \text{joules per bit} \qquad 9.12$$

Repeating the calculation for $S/N = 5$, 0.5, and 0.05, the cost per bit works out at $1.94\,kT$, $0.857kT$, and $0.710kT$, respectively.

The reader may object that, as $S/N \to 0$, the channel capacity will also tend to zero. To see that this need not happen, consider the expression for C, still assuming that the noise is white noise at temperature T so that $N = kTF$

$$C = F \log_2 \left(1 + \frac{S}{kTF}\right)$$

$$= \frac{F}{0.693} \log_e \left(1 + \frac{S}{kTF}\right)$$

Minimum energy per bit is available when $S \ll kTF$ whereupon

$$C = \frac{S}{0.693\,kT} \qquad 9.13$$

and it is seen that the channel capacity may be set at any desired value by choosing S according to this last relation, and making F sufficiently large to ensure $S \ll kTF$. A communication system like this is purely hypothetical, just as a reversible heat engine working in a Carnot cycle is hypothetical, but it tells us the best that is possible. When working against white noise, an 'ideal' communication system can send error-free messages at minimum energy cost per bit, by working at a very small signal-to-noise ratio and with very wide bandwidth. In the case of thermal noise at temperature T, the least cost is $0.693kT$ joules per bit. An interesting comparison of modulation systems from this point of view is given by Raisbeck (see the suggested further reading list at the end of this chapter).

By way of an exercise to illustrate this last result, let us imagine a space station 40 million miles away, which is required to send error-free English text back to Earth at 100 words per minute. What is a rough estimate of the minimum transmitter power needed, if the sky noise temperature in the direction of the space station is $10°K$ at all frequencies?

For simplicity we will assume the transmitting aerial radiates isotropically. The receiving aerial might be a parabolic dish 20 m in diameter, which would subtend a solid angle of $7.2 . 10^{-20}$ steradians at the transmitter and so collect a fraction $6 . 10^{-21}$ of the radiated power. The received power must exceed $0.693kTC$ W, where C is the capacity needed for 100 words per minute. Taking 5 bits per word as the entropy of English text, we have $C = 8\frac{1}{3}$ bits per second, so that the minimum received power is $0.693kT \times 8\frac{1}{3} = 8 . 10^{-22}$ W. The radiated power must therefore exceed 0.13 W. This is astonishingly small and assumes the three unattainables of ideal coding, infinite bandwidth and noiseless amplifiers.

Information transmission from this distance was first achieved in the spectacular flight of Mariner II to Venus in 1962. Scientific and engineering data were time-multiplexed in the spacecraft to yield a p.c.m. signal with $8\frac{1}{3}$ binary symbols per second. These were used to produce a binary phase-shift-keyed subcarrier, which in turn phase-modulated the main carrier at 960 MHz. The radiated power was 3 W and, by using high-gain aerials, the received power was in excess of 10^{-19} W against a sky noise temperature of $10°K$ (the overall noise temperature, using a liquid-helium cooled ruby maser, was approximately $33°K$). This system is clearly beginning to approach the optimum in joules per bit. An order of magnitude improvement would appear unlikely, although more complex coding schemes are contemplated and we may look for yet more exciting developments in this new and highly challenging field.

FURTHER READING

SHANNON, C. E., 'Communication in the presence of noise', *Proc. Inst. Radio Engrs* **37**, 10–21 (1949)

SHANNON, C. E., 'A mathematical theory of communication', *Bell Syst. tech. J.* **27**, 379 and 623 (1948)

SHANNON, C. E., and WEAVER, W., *The Mathematical Theory of Communication* (University of Illinois Press, Urbana, 1949)

Signals and Information

SHANNON, C. E., 'Prediction and entropy of printed english', *Bell Syst. tech. J.* **3**, 50 (1951)

HAMMING, R. W., 'Error-detecting and error-correcting codes', *Bell Syst. tech. J.* **29**, 147 (1950)

RAISBECK, G., *Information Theory, an Introduction for Scientists and Engineers* (Cambridge, M. I. T. Press 1963)

STIFFLER, J. J., *Space Technology*, vol. V, 'Telecommunications' (Scientific and Technical Information Division, NASA, Washington DC, 1966)

The Functions $\text{sinc}(x) = \dfrac{\sin x}{x}$ and $Si(x) = \displaystyle\int_0^x \text{sinc}\, u\, du$

x	sinc (x)	$Si(x)$	x	sinc (x)	$Si(x)$
0·2	0·993	0·200	4·6	− 0·216	1·633
0·4	0·974	0·397	4·8	− 0·208	1·590
0·6	0·941	0·588	5·0	− 0·192	1·550
0·8	0·897	0·772	5·2	− 0·170	1·514
1·0	0·842	0·946	5·4	− 0·143	1·482
1·2	0·777	1·108	5·6	− 0·113	1·457
1·4	0·704	1·256	5·8	− 0·080	1·437
$\pi/2$	0·637	1·370	6·0	− 0·047	1·425
1·6	0·625	1·389	6·2	− 0·013	1·419
1·8	0·541	1·506	2π	0·000	1·419
(1·896)	(0·500)	(1·555)	6·4	0·018	1·419
2·0	0·455	1·605	6·6	0·907	1·426
2·2	0·368	1·688	6·8	0·073	1·438
2·4	0.281	1·753	7·0	0·094	1·455
2·6	0·198	1·800	7·2	0·110	1·475
2·8	0·120	1·832	7·4	0·121	1·498
3·0	0·047	1·849	7·6	0·127	1·523
π	0·000	1·851	7·8	0·128	1·549
3·2	− 0·018	1·851	8·0	0·124	1·574
3·4	− 0·075	1·842	8·2	0·115	1·598
3·6	− 0·123	1·822	8·4	0·102	1·620
3·8	− 0·161	1·793	8·6	0·085	1·639
4·0	− 0·189	1·758	8·8	0·067	1·654
4·2	− 0·208	1·718	9·0	0·046	1·665
4·4	− 0·216	1·676	9·2	0·024	1·672

x	sinc (x)	$Si(x)$	x	sinc (x)	$Si(x)$
9·4	0·003	1·675	11·2	− 0·087	1·560
3π	0·000	1·675	11·4	− 0·081	1·544
9·6	− 0·018	1·673	11·6	− 0·071	1·528
9·8	− 0·037	1·668	11·8	− 0·059	1·515
10·0	− 0·054	1·658	12·0	− 0·045	1·505
10·2	− 0·069	1·646	12·2	− 0·029	1·498
10·4	− 0·080	1·631	12·4	− 0·013	1·493
10·6	− 0·087	1·614	4π	0·000	1·492
10·8	− 0·091	1·597	∞	0·000	1·571 $(\pi/2)$
11·0	− 0·091	1·578			

Appendix 2

Bessel Functions of Order *n*

(Values less than 0·001 omitted)

β	$J_0(\beta)$	$J_1(\beta)$	$J_2(\beta)$	$J_3(\beta)$	$J_4(\beta)$	$J_5(\beta)$	$J_6(\beta)$	$J_7(\beta)$	$J_8(\beta)$	$J_9(\beta)$	$J_{10}(\beta)$
0·2	+0·990	+0·099	+0·005	—	—	—	—	—	—	—	—
0·4	+0·960	+0·196	+0·020	+0·001	—	—	—	—	—	—	—
0·6	+0·912	+0·287	+0·044	+0·004	—	—	—	—	—	—	—
0·8	+0·846	+0·369	+0·076	+0·010	+0·001	—	—	—	—	—	—
1·0	+0·765	+0·440	+0·115	+0·020	+0·002	—	—	—	—	—	—
1·2	+0·671	+0·498	+0·159	+0·033	+0·005	—	—	—	—	—	—
1·4	+0·567	+0·542	+0·207	+0·050	+0·009	+0·001	—	—	—	—	—
1·6	+0·455	+0·570	+0·257	+0·073	+0·015	+0·002	—	—	—	—	—
1·8	+0·340	+0·582	+0·306	+0·099	+0·023	+0·004	—	—	—	—	—
2·0	+0·224	+0·577	+0·353	+0·129	+0·034	+0·007	+0·001	—	—	—	—
2·5	−0·048	+0·497	+0·446	+0·217	+0·074	+0·020	+0·004	—	—	—	—
3·0	−0·260	+0·339	+0·486	+0·309	+0·132	+0·043	+0·011	+0·003	—	—	—
3·5	−0·380	+0·137	+0·459	+0·387	+0·204	+0·080	+0·025	+0·007	+0·002	—	—
4	−0·397	−0·066	+0·364	+0·430	+0·281	+0·132	+0·049	+0·015	+0·004	—	—
4·5	−0·321	−0·231	+0·218	+0·425	+0·348	+0·195	+0·084	+0·030	+0·009	+0·002	—
5	−0·178	−0·328	+0·047	+0·365	+0·391	+0·261	+0·131	+0·053	+0·018	+0·006	+0·001
6	+0·151	−0·277	−0·243	+0·115	+0·358	+0·362	+0·246	+0·130	+0·057	+0·021	+0·007
7	+0·300	−0·005	−0·301	−0·168	+0·158	+0·348	+0·339	+0·234	+0·128	+0·059	+0·024
8	+0·172	+0·235	−0·113	−0·291	−0·105	+0·186	+0·338	+0·321	+0·224	+0·126	+0·061
9	−0·090	+0·245	+0·145	−0·181	−0·266	−0·055	+0·204	+0·328	+0·305	+0·215	+0·125
10	−0·245	+0·046	+0·255	+0·058	−0·220	−0·234	−0·014	+0·217	+0·318	+0·292	+0·207

Vector Representation of Signals

Consider the functions $u_k(t)$, where k is zero or a positive integer, defined as

$$u_0(t) = \frac{1}{\sqrt{T}}$$

$$\left. \begin{aligned} u_k(t) &= \sqrt{\left(\frac{2}{T}\right)} \cos \frac{(k+1)\omega_1 t}{2} \quad && k \text{ odd} \\ &= \sqrt{\left(\frac{2}{T}\right)} \sin \frac{k\omega_1 t}{2} \quad && k \text{ even} \end{aligned} \right\} \qquad 1$$

where $\omega_1 = 2\pi/T$.

These functions are essentially the sine or cosine functions of the Fourier series expansion multiplied by a factor which has been chosen to make

$$\left. \begin{aligned} \int_{-T/2}^{T/2} u_k(t)\, u_l(t) \, \mathrm{d}t &= 0 \quad && k \neq l \\ &= 1 \quad && k = l \end{aligned} \right\} \qquad 2$$

i.e., the functions are said to be 'normalised'. Equations 2.3 and 2.4 may now be written in the rather tidier form

$$v(t) = \sum_{k=0}^{\infty} a_k u_k(t) \qquad 3$$

where

$$a_k = \int_{-T/2}^{T/2} v(t)\, u_k(t) \, \mathrm{d}t \qquad 4$$

We restrict the discussion to signals having no Fourier components above some arbitrarily chosen frequency limit, $a_k \equiv 0$ for $k > K$. We may now imagine a cartesian reference frame in a hyperspace having $K + 1$ dimensions and construct the vector, $\mathbf{V}$, extending from the origin to the point whose coordinates are $(a_0, a_1, a_2, \ldots a_K)$. This (stationary) vector may be used to represent the signal $v(t)$, for its coordinates carry all the information needed to describe the signal. Similarly the vector $\mathbf{W}$, with components $(b_0, b_1, b_2, \ldots b_K)$ represents a second signal

$$w(t) = \sum_{k=0}^{\infty} b_k u_k(t)$$

The elementary signals $u_k(t)$ are represented as unit vectors along the coordinate axes.

The vector which represents the signal $v(t) + w(t)$ has, by equation 4, components $(a_k + b_k)$ but these are the components of the vector sum $\mathbf{V} + \mathbf{W}$. It follows that signal vectors add and subtract like ordinary vectors. An interpretation of the scalar or inner product of two vectors follows from examining the integral

$$\int_{-T/2}^{T/2} v(t)\, w(t)\, \mathrm{d}t = \int_{-T/2}^{T/2} \left(\sum_k a_k u_k(t) \right) \left(\sum_l b_l u_l(t) \right) \mathrm{d}t$$
$$= a_0 b_0 + a_1 b_1 + a_2 b_2 + \ldots + a_k b_k \qquad 5$$

The right-hand side is the familiar expression for the inner product $\mathbf{V}.\mathbf{W}$. Hence the inner product of two signals is given by the integral on the left-hand side of equation 5.

If two vectors are orthogonal their inner product vanishes (Section 2.7). The elementary vectors $\mathbf{u}_k$ are mutually orthogonal, lying along the cartesian axes, in agreement with equation 2. The inner product of a signal with itself represents the total energy in the signal, which is associated with the square of length of the vector in hyperspace

$$E = \int_{-T/2}^{T/2} v^2(t)\, \mathrm{d}t = |\mathbf{V}|^2 \qquad 6$$

The vector representation has been illustrated here using the Fourier components as component vectors. Other sets of functions exist which obey equations 2 and which enable expansions of the form 3 and 4 to be carried out. Two further examples are the $\mathrm{e}^{jk\omega_1 t}$

293

function of the complex Fourier expansion and the sinc $W(t-k/\pi)$ functions of the sampling theorem. A signal space which uses the sample values as coordinates is used in Chapter 9. Further discussion may be found in the article by C. E. Shannon, 'Communications in the presence of noise', *Proc. Inst. Radio Engrs* **37**, 10 (1949).

Autocorrelation Function of White Gaussian Noise

A typical term in the expansion of equation 7.3 is

$$R_{kl} = \frac{1}{T} n_k n_l \int_{-T/2}^{T/2} \text{sinc } W \left(t - \frac{k\pi}{W} \right) \text{sinc } W \left(t + s - \frac{l\pi}{W} \right) dt$$

By making T sufficiently large, this integral may be made as close as we please to the same integral with infinite limits and then transformed with the aid of equation 2.43 to give

$$R_{kl} = \frac{n_k n_l}{2\pi T} \int_{-\infty}^{\infty} G_k(\omega) \, [G_l(\omega) \, e^{j\omega s}]^* \, d\omega$$

where

$$G_k(\omega) = \frac{\pi}{W} e^{-jk\pi\omega/W} \qquad -W < \omega < W$$

$$= 0 \qquad\qquad |\omega| > W$$

is the Fourier transform of sinc $W(t - k\pi/W)$, hence

$$R_{kl} = \frac{n_k n_l}{2\pi T} \left(\frac{\pi}{W} \right)^2 \int_{-W}^{W} e^{-j\omega[(k-l)\pi/W + s]} \, d\omega$$

$$= \frac{n_k n_l}{2\pi T} \left(\frac{\pi}{W} \right)^2 . 2W \text{ sinc } [Ws + (k-l)\pi]$$

This has next to be summed over all k, l. A partial sum over terms for which $k - l = m$, where m is any integer other than zero, yields a contribution

$$\text{const.} \times \sum_{k-l=m} n_k n_l$$

but as n_k and n_l are independent Gaussian variables of zero mean the mean of their product is zero. If we now average over the ensemble, this contribution vanishes and we are left with those terms for which $k = l$; this gives

$$R(s) = \frac{\pi}{WT} \text{ sinc } Ws \sum_k n_k^2$$

There are WT/π terms in the sum, and the mean of n_k^2 is σ^2, hence we have

$$R(s) = \sigma^2 \text{ sinc } Ws \qquad\qquad 7.4$$

for the ensemble average of the autocorrelation function for band-limited white Gaussian noise.

Fourier Analysis of the Noise Signal of Section 7.4

A Gaussian noise signal, band limited to W rad s^{-1}, is sampled in the range $t = -T/2$ to $t = T/2$ at the moments $t = k\pi/W$. From these samples a repetitive waveform is imagined, such that the samples of this new waveform at the moments $t = k\pi/W$, $k\pi/W \pm T$, $k\pi/W \pm 2T$, etc., are identical. The α-coefficient at angular frequency $m\omega_1$ of this repetitive waveform must be given by the linear superposition

$$\alpha(m\omega_1) = \sum_k \alpha_k$$

where α_k is the α-coefficient at $m\omega_1$ due to that part of the waveform arising from the samples, each equal to n_k, at $t = k\pi/W$, $k\pi/W \pm T$, etc. By the sampling theorem this part of the waveform is a train of pulses each of the form n_k sinc Wt with period T. The spectrum of such a pulse train was investigated in Section 2.17 from which we have immediately that

$$\alpha_0 = \frac{\pi n_0}{WT}.$$

The contribution to $\alpha(m\omega_1)$ due to the sample n_k at $t = k\pi/W$ and its partners suffers a phase shift arising from the time delay $k\pi/W$, thus

$$\alpha_k = \frac{\pi n_k}{WT} \, e^{-jkm\pi\omega_1/W}$$

Hence

$$\alpha(m\omega_1) = \frac{\pi}{WT} \sum_k n_k \, e^{-jkm\pi\omega_1/W} \qquad\qquad 1$$

The real part is

$$\mathcal{R}\alpha(m\omega_1) = \frac{\pi}{WT} \sum_k n_k \cos k\theta$$

where $\qquad \theta = m\pi\omega_1/W \quad$ and thus $\quad -\pi < \theta < \pi$

Once more we find that we are dealing with a linear superposition of independent Gaussian variables, n_k, so that $\mathcal{R}\alpha(m\omega_1)$ is itself a Gaussian random variable. That is to say, working across the ensemble and picking out this quantity for each sample function we would obtain a sequence of normally distributed numbers. Averaging across the ensemble and using equations 5.23 and 5.24 we find

$$E[\mathcal{R}\alpha(m\omega_1)] = 0$$

and

$$\text{Var}\,[\mathcal{R}\alpha(m\omega_1)] = \frac{\pi^2 \overline{n^2}}{W^2 T^2} \sum_{k=-WT/2\pi}^{WT/2\pi} \cos^2 k\theta$$

There are WT/π terms in the summation and, if T is sufficiently large, the sum may be replaced as closely as we please by $WT/2\pi$. Hence, equating variance to mean square as the mean is zero

$$\overline{[\mathcal{R}\alpha(m\omega_1)]^2} = \frac{\pi \overline{n^2}}{2WT} = \frac{p_n\omega_1}{4}$$

Similarly for the imaginary part of $\alpha(m\omega_1)$

$$\overline{[\mathcal{J}\alpha(m\omega_1)]^2} = \frac{p_n\omega_1}{4}$$

It may also shown that the expectation of the product $\mathcal{R}\alpha(m\omega_1) \times \mathcal{J}\alpha(m\omega_1)$, averaged across the ensemble, is zero. These quantities are thus uncorrelated and, being Gaussian, are independent.

The mean square of the modulus of the α-coefficients is found by adding the mean squares of the real and imaginary parts, thus

$$\overline{|\alpha(m\omega_1)|^2} = \frac{p_n\omega_1}{2}$$

There is no angular preference in equation 1, so the phase angle of α is quite randomly distributed around 2π (but note that $\alpha(-m\omega_1) = \alpha^*(m\omega_1)$ as is always the case). Equation 7.8 shows that the noise power is independent of the frequency, $m\omega_1$, *after averaging across the ensemble*. For a particular sample function we have no right to expect all the α-coefficients to have the same magnitude, but if T is sufficiently large, then any practical measurement of noise power will collect power from the large number of coefficients embraced by the system bandwidth. This averaging process means essentially that the power density is uniform, as expected.

The Fourier sine and cosine coefficients are found by combining $\alpha(m\omega_1)$ and $\alpha(-m\omega_1)$ so that we may write

$$n(t) = \sum_{m=1}^{WT/2\pi} a_m \cos m\omega_1 t - \sum_{m=1}^{WT/2\pi} b_m \sin m\omega_1 t \qquad 2$$

where
$$a_m = 2\mathcal{R}\alpha(m\omega_1)$$
$$b_m = 2\mathcal{J}\alpha(m\omega_1)$$

showing that a_m and b_m are also (independent) random Gaussian variables of zero mean. Their mean square value is

$$\overline{a_m^2} = \overline{b_m^2} = p_n\omega_1 \qquad 3$$

In equation 2 the d.c. term has for convenience been omitted. From equation 1 this has mean square value

$$\overline{a_0^2} = \frac{p_n\omega_1}{2}$$
$$= \frac{\overline{n^2}}{N} \qquad 4$$

where N is the total number of independent samples in the time interval.

Index

Index

Index